BUILD A REMOTE-CONTROLLED ROBOT

Other TAB Electronics Robotics Titles

The Robot Builder's Bonanza, Second Edition, by Gordon McComb

Robots, Androids, and Animatrons, Second Edition, by John Iovine

TAB Electronics Build Your Own Robot Kit by Myke Predko and Ben Wirz

BUILD A REMOTE-CONTROLLED ROBOT

DAVID R. SHIRCLIFF

McGraw-Hill
New York • Chicago • San Francisco • Lisbon • London • Madrid
Mexico City • Milan • New Delhi • San Juan • Seoul
Singapore • Sydney • Toronto

Cataloging-in-Publication Data is on file with the Library of Congress

McGraw-Hill

A Division of The McGraw·Hill Companies

1 2 3 4 5 6 7 8 9 0   DOC/DOC   0 8 7 6 5 4 3 2

ISBN 0-07-138543-6

The sponsoring editor for this book was Scott L. Grillo, the editing supervisor was Stephen M. Smith, and the production supervisor was Sherri Souffrance. It was set in Fairfield Medium following the NBF design by Deirdre Sheean and Vicki Hunt of McGraw-Hill Professional's Hightstown, N.J., composition unit.

Printed and bound by R. R. Donnelley & Sons Company.

Previously published as *Build a Remote-Controlled Robot for under $300*, copyright © 1986 by TAB Books, Inc.

This book is printed on recycled, acid-free paper containing a minimum of 50% recycled, de-inked fiber.

To my wife, Raye, and my children, Daniel, Haley, Ian, Margaret, and Raymond, for their support and encouragement.

ABOUT THE AUTHOR

David Shircliff is a teacher at Seneca Ridge Middle School in Loudon County, Virginia, where he teaches classes in technology education. A dedicated electronics enthusiast, Mr. Shircliff has been researching and building robots for over 20 years.

CONTENTS

PREFACE

In recent years robots have captured the interest of more and more people. Thanks to movies and TV, the notion of the robot as a mechanical companion and servant has become a common concept. As interest in robots grew, a number of books showing how to build robots at home began to appear. These books, however, were very technical, showing how to build computer-controlled mobile platforms that are considered by most to be true robots.

My interest in robots leaned more toward the popular concept of robots as humanlike friends and servants. I did not have the technical skill or funds to build a computer-controlled robot, so I decided to develop a robot that would fit the popular image of robots and not be too difficult to complete or expensive to build. The result was Questor.

While working on Questor, I tried to develop a project that I, as a beginner, could complete with little technical skill, using tools I had in my workshop. Also, I wanted Questor to look and function like a robot butler, a form I felt best fit the friend/servant theme. For this reason I needed a people-sized robot that would have great presence. I concentrated more on form than sophistication to develop an impressive looking, but relatively simple-to-build, project—a beginner's project.

Later, when I decided to write a book about the project, I wanted to avoid weaknesses I found in other how-to robot books. This book is heavily illustrated, helping to take the guesswork out of Questor's construction. Next, the book deals only with the construction of the robot, and not the theories on which it is based. This type of information is best derived from specialty electronics and robotics books. I have included a list of books and magazines that supply information, as well as other possible sources for robot kits and parts.

It is my hope that you will use this book not only to build your own version of Questor, but to guide you in creating your own unique robot. This way your robot will reflect your knowledge and skill as a builder. Also, I hope that your robot will be used as a test bed for other robotics projects. If you are like me, once you build your own robot, you'll always be trying to improve it.

David R. Shircliff

INTRODUCTION

One of the first questions you will have to answer when you say you have your own robot is, "What does it do?" If your answer (as mine) is, "It rolls around by remote control and serves drinks" disappoints the questioner, don't be offended. It simply means that the person asking the question knows little about the real world of robotics, the science of robots.

Before you can attempt to explain your answer to the uninformed asker, you must know a little about the subject of robots. Ask yourself, "What is a robot?" The word robot comes from the Czech word *Robota*, which means obligatory work or servitude. The word robot was first used in a Czech play called *R.U.R. (Rossum's Universal Robots)* by Karl Capek. Written in 1921, the play depicts a race of humanoid robots that turn on their masters and destroy them, a theme that seems always to be associated with robots. Figure I-1 shows a scene from the play.

The exact meaning of the term robot, even in today's technological age, is a matter of debate. Man's technical prowess makes the exact meaning elusive: manlike mechanical device; person working mechanically, without original thought; machine or device that works automatically. These definitions seem rather broad and could encompass any number of modern devices from a dishwasher to a timer-controlled video cassette recorder, without conjuring up the popular *Star Wars* notion of robots.

A second, more-precise definition is stated by the Robot Institute of America. It reads: "A robot is a programmable multifunctional manipulator designed to move material, parts, tools or specialized devices through variable programmed motions for the performance of a variety of tasks."

While more precise, it tends to be narrow and also does not parallel the popular notion of the mechanical friend everyone

FIGURE I-1.   The robots of the play *R.U.R.* (*Rossum's Universal Robots*) attack their human masters. (*Courtesy of New York Public Library at Lincoln Center.*)

would like to have. It applies more specifically to those types of robots at work in factories all over the world, shown here in Figs. I-2 through I-4. These assembly line type robots can do everything from welding a car (then painting it) to assembling delicate electronics components, all automatically, 24 hours a day if needed, and without a break. They don't get sick (although when they do break down, they can be easily repaired or even replaced), ask for pay raises, or any pay for that matter, and can be retrained to do another job in a matter of minutes by simply changing the job program in their control computers. If you look again at Figs. I-2 through I-4, you will see that while the device most certainly looks mechanical, it does not look like a human. Instead it takes the shape of the most useful part of the human anatomy, from a robot standpoint, the arm.

Both these definitions seem to be correct in their specific case, but there is a middle family between the simple automated device and the sophisticated computer-controlled

FIGURE I-2.   An industrial robot. (*Courtesy of Cincinnati Milacron.*)

manipulator. This middle family is that of the show robot or showbot. Questor, the robot outlined in this book, is a member of the showbot family. Figures I-5 through I-8 picture examples of commercial show robots.

A showbot in most cases has no computer brain. Instead it is controlled via a remote control system operated by a person somewhere out of sight. You might have seen or heard of a

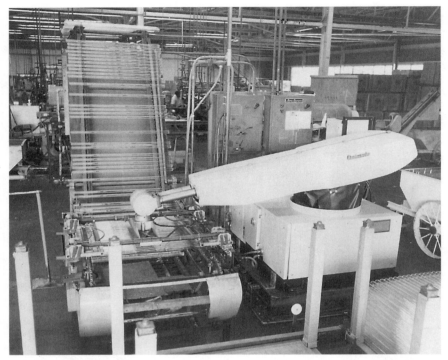

FIGURE I-3.   Robots are best used for repetitive tasks like stacking. (*Courtesy of Unimation Inc.*)

FIGURE I-4.   Robots can do light work such as grinding. (*Courtesy of Cincinnati Milacron.*)

FIGURE I-5.   Showbots come in many shapes and sizes. (*Courtesy of Ken Zaken, Robots 4 Fun.*)

showbot entertaining groups of fascinated people in shopping malls or on TV as characters in movies. I even read about a showbot delivering a speech at a college graduation. Showbots, however, can be adapted for use in the home.

This book lays the groundwork to construct one such home showbot, Questor. (See Figs. I-9 and I-10.) Questor was designed to look like and function like a butler. There is a drink dispenser built into his arm and a vacuum port in his mobile platform. I felt these two functions are what most people expect a robot servant to do. The arms, which help promote Questor's humanoid shape, are nonfunctional; they serve only to hold the serving tray. The hands are made of two auto drink holders. A button located on the wrist (the area above where the hands are bolted on) controls the drink dispenser.

His head is a lamp, and there are two headlights on the front of the mobile platform. These lights not only help the operator guide the showbot at night, but they are very useful during power blackouts. There is also a 12-volt direct current (dc) cigarette lighter plug on the side of the base. This is used to run battery-powered appliances such as portable radios or TVs off the robot's batteries.

FIGURE I-6.   The Six-T showbot can blow up
balloons! (*Courtesy of The Robot Factory.*)

A horn located on the lower part of the front body panel
announces Questor's presence. I plan to add a tape recorder for
prerecorded messages. This is something you could consider
designing into your showbot. Finally, his body panels and arms
were painted to look as though Questor is wearing a tuxedo
jacket, and a light-up bow tie completes the look.

I also designed Questor so he could be built using tools
found in a home workshop and parts available in local hardware and electronics stores. However, there are a few parts
you will have to order. The following list of what I've deter-
mined are "must buy parts" shows items you will need to purchase before starting construction. The address for a parts
supplier, Herbach & Rademan Company, is listed in Sources
in the back of the book.

FIGURE I-7. Showbots can also be soft and fuzzy. (*Courtesy of The Robot Factory.*)

Must Buy Parts

2 12-volt dc motorized wheels

2 6-volt, 8-amp solid gel batteries, with charge kit

2 10-ohm, 25-watt potentiometers

Note: The drink dispenser motor and vacuum system kit can also be ordered from Herbach & Rademan. The rest of the parts needed for each phase of Questor's construction will be listed in the beginning of each chapter.

All of Questor's various components, except for the remote control system, are powered by a 12-volt dc battery system. Questor can be controlled by either a control box connected to the base by a cable or a wireless remote control system. The

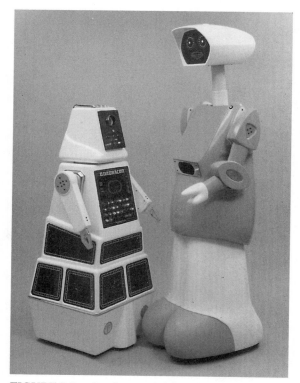

FIGURE I-8. Quadracon and friend Peeper. (*Courtesy of Pelican Beach LLC, successor to ShowAmerica Inc.*)

remote control system, as are the rest of the parts, is a standard off-the-shelf item.

At this point, you should read through the book to familiarize yourself with the diagrams, photographs, parts lists, and overall format. Once you plan your showbot, you can order the "must buy" items. You are now ready to enter the fascinating world of robotics.

ROBOT BASICS

But first, a review of the basics.

The construction of a remote-controlled robot, while not easy, need not be difficult. My motto when designing and building Questor was "keep it simple, stupid" (KISS)! The

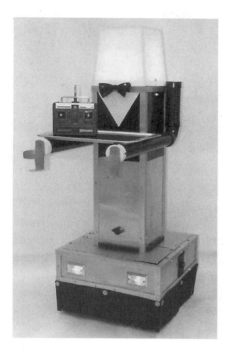

FIGURE I-9.   Questor the
robot servant (front view).

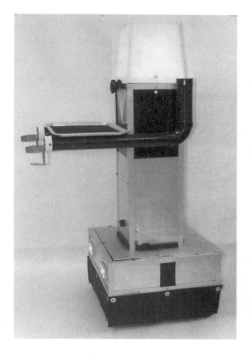

FIGURE I-10.   Side view.

best strategy is to use as many off-the-shelf items as possible. As your confidence and skill level grow, you can design and build your own components. In addition, as you work with different materials, such as wood, plastic, and metal, you will learn the properties of each and how you can use them in your own robot designs.

When I first started to design and build remote-controlled robots I used a very simple motorized wheel assembly taken out of a toy car and made bodies out of poster board and construction paper. Figure I-11 shows the plans for one such robot. I tried to make these robots life size, 3 to 4 feet tall. They were fun to design and build and taught me a lot about what would work without being expensive to construct. If you are a first-time robot builder, I suggest that you try one of these paper robots. Whether made of paper or wood and metal, all my robot designs have four basic subsystems: a motorized base, a remote control system, a power supply, and a body.

MOTORIZED BASE

The motorized base for your robot can be the most difficult subsystem to design and build. You can save yourself a lot of trouble if you design "around" this part of the robot. Instead of designing the robot first and fitting the motorized base to your design, design and build the base first and then fit your robot body to it. The base generally holds all of the internal parts or "guts" that make your robot work. In Questor, for example, the base has the wheels mounted on it as well as the batteries (which can be quite heavy) and the control system. That is why he looks the way he does. The old saying "form follows function" is true in robot design, too.

The most important part of the base is the motor-driven wheels. This is where many (myself included) robot builders have the most problems. Do yourself a favor and buy motorized wheel units. A supplier is listed in the back of the book. These units already have a motor mounted to a drive wheel and usually the assembly is in a frame you can modify to attach to your base. With Questor's motorized wheels all I had to do was design a way to mount them to the base. Figure I-12

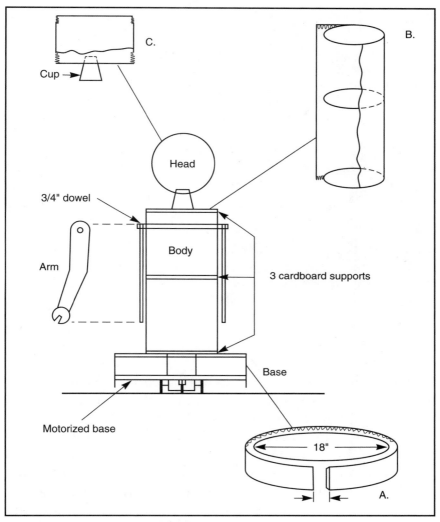

FIGURE I-11.   Complete plans for a "paperbot."

shows the wheels I used for Questor. Another good source for motorized wheels is children's ride-on toys. They already have the motors and wheels mounted to a frame (as well as the batteries) and can be used as is, or removed and mounted on your robot base. The main drawback of using ride-on toys is they lack a steering system. This is a critical area that you must design into your base.

FIGURE I-12.   Motorized wheel.

There are several ways to set up a steering system for your robot. A robot must have a minimum of three wheels in order to work, and how you power and mount the wheels will affect your control of the robot. Figure I-13 shows the combinations available. All the combinations require two motorized wheels and at least one swiveling wheel for balance. This was the system I used in Questor. By making one motorized wheel go forward or in reverse, while the other is off or going forward, you can very effectively steer a robot. Table I-1 charts the combinations for steering with this system. All this "control" is provided by the next subsystem, the remote control system.

REMOTE CONTROL SYSTEM

The ability to remotely control your robot is a big part of its appeal. The two types of remote control are *wired* and *wireless*. Basically, what you are doing with either system is tripping switches to control robot functions. Figure I-14 shows the two systems, as used for Questor.

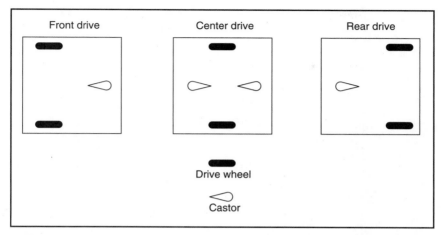

FIGURE I-13. Drive wheel layouts.

TABLE **I-1.** Control Combinations

MOTORIZED WHEEL ONE	MOTORIZED WHEEL TWO	ROBOT DIRECTION
On	On	Forward
Off	Off	Stop
On	Off	Right turn
Off	On	Left turn
On	Reverse	Spin right
Reverse	On	Spin left

With a wired remote control system the operator uses a control box connected to the robot via a long wire or cable. The advantages of this system are that it is simple to build and costs little to construct. The main disadvantages are limited range and the cable itself can get in the way. For the beginning robot builder, however, this is the best system to start out with. It will allow you to build and test systems for your robot without the complexity and expense of a wireless system.

Wireless remote control is what most people think of when you say "remote control." A wireless system allows a much greater range for the operator, and there is no control cable to get in the way. A wireless system has three main parts. The

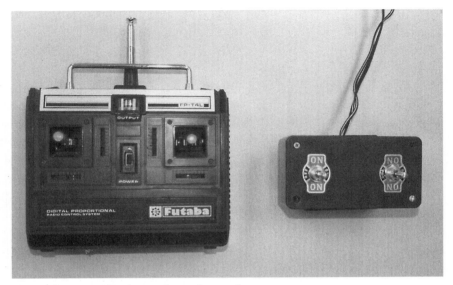

FIGURE I-14. Wireless and wired controls.

first is a transmitter. This is what you will use to control your robot. The transmitter sends signals to a receiver. The receiver is the second part of the system; it uses the signals to control servos, the third part of the system. A servo takes decoded signals from the receiver and uses them to turn a motor. Figure I-15 shows a complete wireless system. Systems like this one are readily available from your local hobby shop. The motor in a servo is very powerful for its size and can be used to trip switches within your robot and control it. Figure I-16 shows the servo/switch control system.

POWER SUPPLY

How to power a robot is another area where the beginning robot builder can experience difficulty. Here my KISS philosophy can again help. You will find, as I did, that different systems have different power needs. The use of rechargeable batteries to power each of these systems is the simplest solution.

Question uses four rechargeable batteries. Two are 6-volt batteries wired together to supply the 12 volts needed to power the drive wheels, main lights, and vacuum cleaner. The

FIGURE I-15. Basic wireless remote control system.

other two batteries are both 9 volts. One powers the receiver used in the remote control system, while the other is used to power the robot's blinking-light bow tie.

The 6-volt batteries came from a motorized children's ride-on toy. They have the advantages of being readily available and being designed to be safe, because they are sealed. Another advantage is that the battery charger and plug needed come with the batteries (Fig. I-17). Currently you can find these batteries in both 6- and 12-volt sizes. The first of the 9-volt batteries came with the remote control system along with a charger for both the receiver and the transmitter. The second 9-volt battery, for the blinking bow tie, is simply a standard rechargeable. Figure I-18 shows the two 9-volt batteries. You will notice that while they are both 9 volts they are different sizes.

By using separate batteries for each system you can avoid having to build the complex circuits needed to raise or lower

FIGURE I-16. Questor's servo/switch control board.

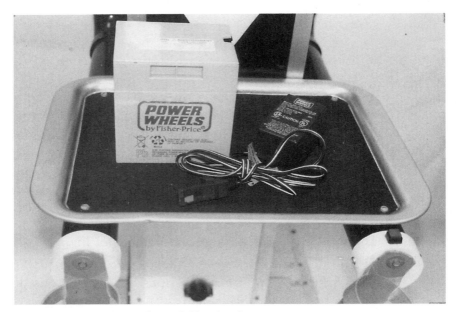

FIGURE I-17. Battery from children's ride-on toy.

FIGURE I-18.   9-volt batteries used in Questor.

battery power. However, if you have many systems on your robot (for example, two or three blinking lights) you will have to recharge a lot of batteries. You can, though, combine systems with the same power needs to feed off the same battery.

Regardless of how many batteries you use, you will have to be able to access them, as well as the other subsystems of your robot. This brings us to the next and, for a showbot, the most important subsystem: the body.

BODY

A robot's body tells the world who the robot is and what it can do. For a show robot the body is its reason for existence, because showbots are used for entertainment and to help advertise products. Questor was designed to be a robot butler. He has a waiter's jacket painted on his body panels, and his arms, while nonfunctional, do hold a serving tray. A blinking bow tie helps complete his servant look. I built his body from scratch, after completing his motorized base, with this specific look in mind.

I am sure that you already have many ideas on how you would like your robot to look. I hope the examples given here will help spark your imagination. You should keep sketches of your robot ideas in a notebook, so as your skill level grows you can attempt your more advanced notions.

One of the key points to keep in mind while designing is, what will you build your robot out of? For the beginning robot builder this can be particularly vital. What materials are you familiar with? Do you have the tools to work with a certain type of material? How much does a material cost? How much will the body weigh when complete? All these are questions that you will have to ask yourself. You can answer most of these questions if you use a body that is already built. Where can you get a prebuilt robot body? At your local variety store they are called trashcans! Before you shudder at the idea that your robot be made out of a trashcan, let us examine the advantages.

The main advantage of using a trashcan is that it is a ready-made container that can be built on. Take for an example a small metal garbage can, like that in Fig. I-19. Being metal (galvanized steel), it is very sturdy and has the added benefit of looking "robotic." Even though it is made of metal it can be easily cut and drilled. (Care should be taken when

FIGURE I-19. Yes, with a little imagination this can be a robot!

FIGURE I-20.   Trashcan-inspired showbot. (*Courtesy of Pelican Beach LLC, successor to ShowAmerica Inc.*)

working with metal because cutting and drilling will produce sharp edges.) The trashcan's size is also of benefit because it helps your design have a "life-sized" look that is important to showbots. Figure I-20 pictures a showbot with a body the shape of a trashcan. If you feel that metal is too difficult to work with, simply buy a plastic trashcan. Be sure that the plastic is hard and nonflexible. Flexible plastic is weak and therefore makes a poor body.

Small metal cans, wash tubs, and even salad bowls can be used for robot bodies. Once your base is done you can experiment with different containers until a final design is reached. Figures I-21 and I-22 on pp. xxx and xxxi show sketches of two robot bodies. Notice how by stacking different containers, two very familiar (and famous) robots come to life.

I hope that this section on robot basics has prepared you for the fun you will encounter in the rest of the book.

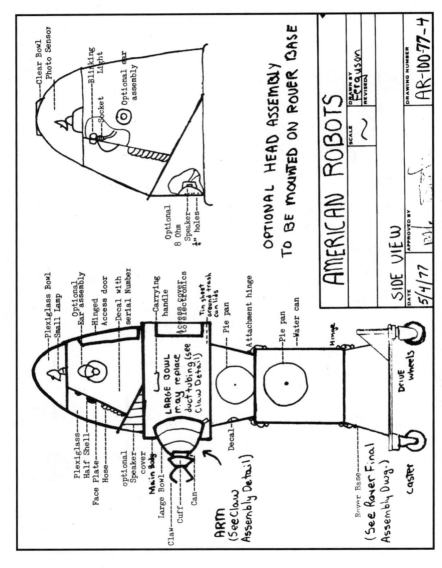

FIGURE I-21. Can you name this robot? (*Courtesy of American Robots.*)

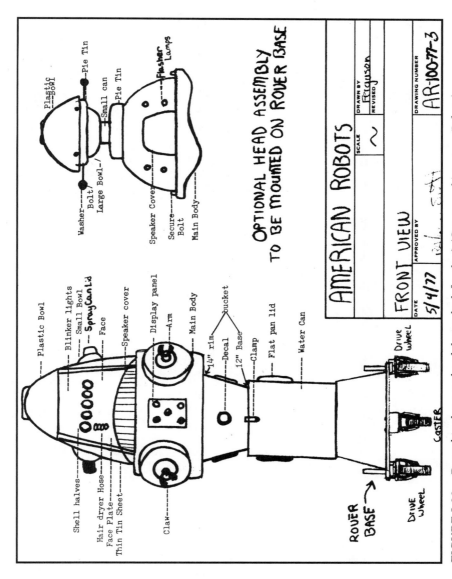

FIGURE I-22. Does this alternate head design look familiar? *(Courtesy of American Robots.)*

THE MOTORIZED PLATFORM

The motorized platform is a most important part of Questor's construction. It not only supplies the robot with mobility, but contributes to its personality and appeal. Although simple in construction, the platform outlined in this chapter is capable of carrying 50 pounds of robot. To start its construction, the first job to perform is to prepare the motorized wheels that propel the platform.

PREPARING MOTORIZED WHEELS

Once you have obtained the motorized wheels, study them and learn how they operate and how they are assembled. This is important because you must disassemble the wheels in order to prepare them for attachment to the platform. Be careful not to lose any of the smaller parts and work on only one wheel at a time. Figures 1-1 and 1-2 show an assembled and disassembled wheel.

To disassemble the wheel, first remove the motor and gearbox assembly held to the side of the wheel's frame by three small screws. On the opposite side of the frame is a cotter pin that holds the wheel's axle and frame together. Remove this pin (this is done easily with a pair of needle nose pliers) and pull the axle out from the other side slowly. As you pull the axle out, four small spacing washers, two red plastic and two metal, will fall from the frame along with the wheel itself and the wheel's large white driver gear.

FIGURE 1-1.   Assembled motorized wheel.

Next remove the swivel ring from the top of the now bare frame. The ring is held in place by a cotter pin that passes through the large post on the top of the frame. Remove this pin and slide the swivel ring up the post and off the frame. There will be some grease and a small ball bearing left on the top of the frame. Wipe away the grease and remove the ball bearing. The swivel ring, cotter pin, and ball bearing are no longer needed for this robot, but add them to your parts supply for later projects.

Now you are ready to prepare the empty frame for attachment to the platform. After considering many different methods of attaching the wheels to the platform, I came to the conclusion that the most direct and simple way is to drill holes in the frame and holes in the platform, then bolt them together. Figure 1-3 shows the location of four 3-inch × 1/4-inch-diameter holes that are to be drilled on the top of the frame. If you have never worked with metal before or do not

FIGURE 1-2.   Disassembled motorized wheel.

have a strong vise, don't attempt to drill the holes yourself.
A local metal shop or school industrial arts class could do it
for you.

Using Fig. 1-3, mark the locations of the holes on the top
of the frame and start the holes with a center punch. This
device makes a small dent in the metal for the drill bit to sit
in. If you don't have a center punch, a nail will do. I found
the easiest way to drill the holes was using a strong vise,
clamp one leg of the frame lengthwise between two pieces of
wood. You will have to bend the legs apart slightly to accomplish this. Now drill the hole marked on the top of the frame
on the side of the leg that you clamped. After drilling,

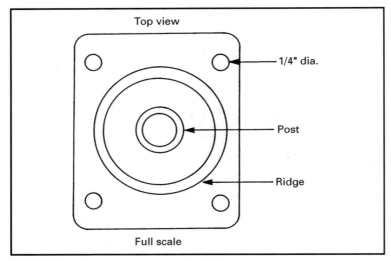

FIGURE 1-3.   Location of mounting holes.

unclamp the leg, flip the frame over, clamp the other leg, and drill the hole on that side. Figure 1-4 shows how to clamp and drill the holes in this way. You could have clamped the frame posts in the vise, but round objects tend to slip when you drill them.

Now reassemble the first motorized wheel and disassemble and drill the second. Figure 1-5 shows the top view of one completed wheel. With both wheels drilled and assembled, it is time to cut and drill the platform.

THE PLATFORM

The platform itself is simply a 20- × 20- × 1/4-inch piece of plywood, cut from a larger 24- × 24- × 1/4-inch piece. While simple in design and construction, it is the key element on which all of Questor is mounted. Great care should be taken to try to keep all of the various holes and cuts as precise and as straight as possible. The easiest way to assure straight cuts is to measure 4 inches in from the bottom edge of one side of the board and 4 inches in from the top edge of the same side, then connect the two points with a line. Figure 1-6 shows how

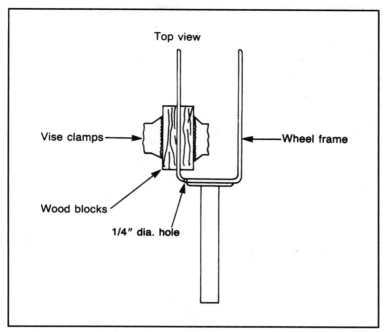

FIGURE 1-4.   Suggested clamping method.

FIGURE 1-5.   Completed motorized wheel.

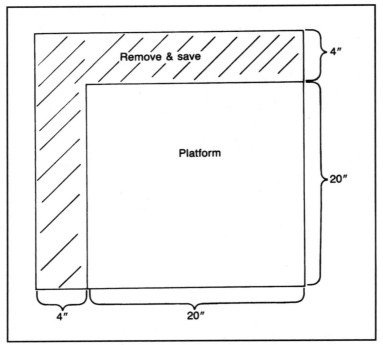

FIGURE 1-6.   Guide for cutting platform; remove and save excess wood.

to mark the plywood for cutting. After you cut the platform from the stock plywood, sand the cut edges to remove any splinters. Save the leftover plywood; it will be used later.

To mount the motorized wheels on the platform you must first drill two 3/4-inch-diameter holes to accommodate the posts of the wheel frame. Figure 1-7 shows where on the platform to drill the holes. If you don't have access to a 3/4-inch-diameter drill bit, you can make the holes by drilling smaller holes around the inside of the 3/4-inch circle, then removing the wood with a coping saw. This rough circle is then filled and sanded to shape.

MOUNTING WHEELS

After the holes are drilled, it is time to mount the motorized wheels. To do this take one of the predrilled motorized

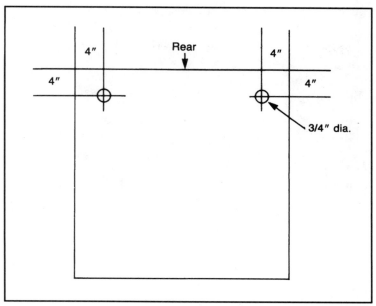

FIGURE 1-7.   Location of post holes.

wheels and insert its frame post into one of the 3/4-inch holes in the platform. Then turn the wheel so that the motor and gearbox face the center of the platform. Figure 1-8 shows the correct position of the wheels. You must make sure that the wheels point as straight as possible during the mounting process.

Each of the motorized wheels is held to the platform by four 2-inch × 1/4-inch-diameter bolts. With the wheel pointing as straight as possible, take a pencil and carefully mark the location of one of the four mounting holes in the wheel's frame on the platform. Remove the motorized wheel and drill a 1/4-inch-diameter hole where marked. Now replace the wheel and realign the hole in the frame with the hole now in the platform. Take one of the eight bolts and insert it through the wheel's frame and out the top of the platform. If the bolt does not go through the hole in the platform easily or after it's inserted the motorized wheel is no longer straight, remove the wheel and redrill the hole in the platform slightly larger. The play in the hole will allow you to shift the wheel's position.

FIGURE 1-8.   Motorized wheels in mounted position.

This, however, only works for small adjustments. With the bolt inserted you can now mark and drill the other holes starting with the hole in the opposite corner from the bolt. (If you started with the bottom left hole, drill the upper right next.) This method and order of marking and drilling helps ensure the wheels will be straight.

The motorized wheels' final attachment to the platform is illustrated in Fig. 1-9. Notice the use of lockwashers. These washers are very important because they keep the nuts from coming loose due to vibration caused when the robot travels over rough surfaces. You should use lockwashers throughout your robot. Also Fig. 1-9 shows the use of a large bore washer. This washer should be approximately 2 inches in diameter with a 3/4-inch bore to allow it to slip over the post of the motorized wheel. The washer keeps a ridge on the top of the wheels' frame from digging into the wooden platform when the bolts are tightened. Also, the washer helps keep the wheel sitting level. After the wheels are attached, make a final check to see that they are straight. Once the two motorized wheels are mounted, it is necessary to mount a third castor wheel on the front of the platform.

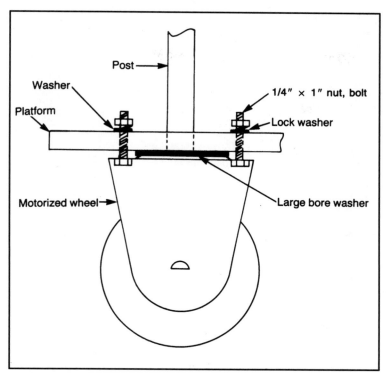

FIGURE 1-9. Motorized wheel attachment.

THIRD CASTOR WHEEL

The castor wheel makes the platform level and gives it stability. The wheel should be about 3 inches in diameter and designed for heavy-duty use. Depending on the wheel you obtain, you'll probably have to stack pieces of plywood between the platform and the mounting plate of the castor wheel so that the platform sits level. Figure 1-10 shows where on the platform the wheel is mounted. Remember to make sure that the castor is centered along the front edge of the platform. A guideline shown in Fig. 1-10 shows how to do this.

After you find the center point, place the castor wheel on the platform so that if turned in a circle, the wheel will not protrude past the front edge of the platform. Next mark and drill the hole for the wheel's mounting plate. (The diameter of the

TABLE 1-1. Parts List

AMOUNT	ITEM
2	Motorized wheel
1	Sheet of 24- × 24- × 1/4-inch plywood
1	4-inch-diameter castor wheel
8	2-inch × 1/4-inch-diameter bolt, nut, and lockwasher set
4	3-inch × 1/4-inch-diameter bolt, nut, and lockwasher set
2	2-inch × 3/4-inch-diameter bore washer
1	Auto fiberglass repair kit, including cloth and resin
1	Can spray paint (color of your choice)

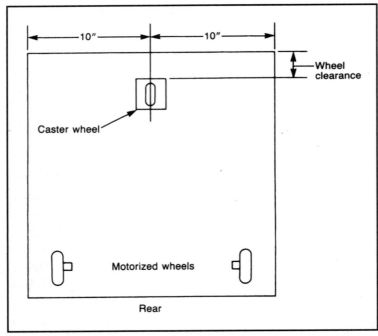

FIGURE 1-10.   Location of castor wheel.

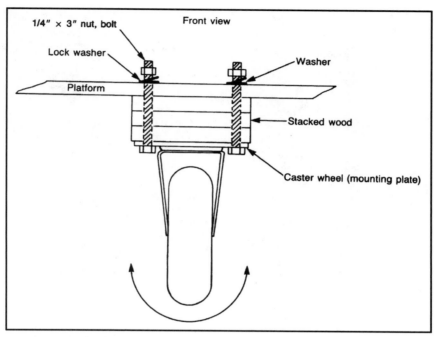

FIGURE 1-11.   Castor wheel attachment.

holes depends on the wheel you have obtained.) As I noted before, you will probably have to stack some plywood spacers (use the wood left over from when you made the platform) between the wheel and the platform. These spacers are made by tracing around the castor's mounting plate and then marking and drilling the mounting holes as you did for the platform. When you stack the spacers, all the holes should line up.

Bolt the castor wheel and spacers to the platform as shown in Fig. 1-11. Then using a small level, check to see that the platform sits correctly. If the level of the platform is slightly off, this can be corrected by placing washers between the stacked plywood and the mounting plate of the castor until level.

FINISHING TOUCHES

After you have mounted all three wheels, remove them and paint the platform. This not only makes the platform look bet-

FIGURE 1-12. Completed platform (bottom view).

ter, but makes it water resistant. You may opt as I have to fiberglass the platform. Fiberglass also provides water protection and adds strength to the platform. Fiberglass is very easy to work with (especially on a flat surface) so if you follow the directions on the package, you should have no problems. If you do use fiberglass, use a kit with a clear resin so you can locate and redrill all the mounting holes in the platform. Once you have fiberglassed and painted, you can reattach the three wheels. Figure 1-12 shows the completed platform.

BODY FRAMEWORK

Questor's body is made from five 8-foot × 1-inch × 1-inch × 1/8-inch strips of aluminum angle. I chose this material over wood or plastic because while slightly more expensive, it is stronger and more lightweight. Also, if care is taken, aluminum is relatively easy to work with. The aluminum angle is used to form two boxes. These boxes are called the upper framework and the lower framework. Once joined, they make up Questor's body.

Before the boxes are constructed each section of aluminum angle used to make up that portion of the body is marked and drilled with holes to be used later in the robot's construction. These predrilled holes are best made when the framework is in pieces rather than when assembled. A chart will list how to assemble each box so all the predrilled holes are in their proper locations when the framework is complete. Remember to take your time and not to cut or drill the aluminum angle until you have checked your measurements or hole locations against the book.

CUTTING ALUMINUM

Figure 2-1 shows how to cut each of the five aluminum strips into the pieces that make up Questor's framework. Cut the strips with a hack saw and use a miter box to achieve straight cuts. Be careful not to cut the aluminum before you have checked your measurements.

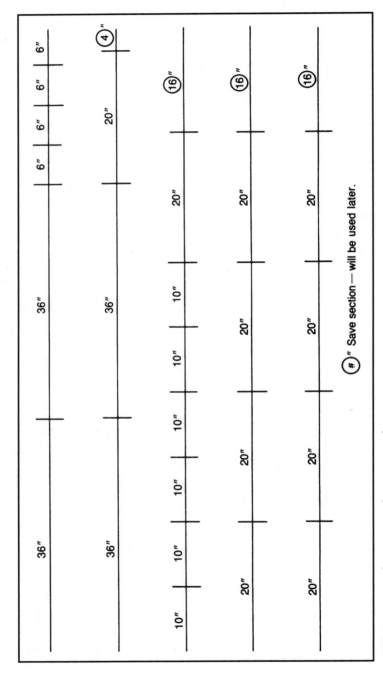

FIGURE 2-1. Aluminum angle cutting guide.

14

TABLE **2-1.** Parts List

AMOUNT	ITEM
5	8-foot × 1-inch × 1-inch × 1/8-inch angle aluminum
50	1/8-inch pop rivet (and rivet gun)
8	1-inch × 5/32-inch-diameter bolt, nut, and lockwasher set
6	1-inch × 1/4-inch-diameter bolt, nut, and lockwasher set
1	Vacuum cleaner kit

Once you have cut the strips into pieces, separate them so that you have four 36-inch, four 6-inch, six 10-inch, and ten 20-inch pieces. Keep all the extra aluminum for use later.

Now that you have the various pieces grouped together, check to see if they are all the same length. If the pieces are slightly unequal, simply choose the shortest piece of that group and cut or file the others down to match it. After all the pieces have been grouped and trimmed, they must be predrilled and some pieces precut before assembly. This preparation will save you a lot of time and trouble later.

DRILLING AND CUTTING THE SECTIONS

Many of the holes to be drilled now are not utilized until later in the robot's construction. It is much easier to drill them now while the framework is in pieces than later when it is assembled. All the cuts to be made consist of 45-degree angles. These cuts are at the ends of one side of some of the pieces and allow them to be joined into squares with no overlap. Figure 2-2 shows an example of this.

Figures 2-3 through 2-26 illustrate how each piece of aluminum is drilled or cut. Each figure consists of two rectangles; one rectangle represents each of the outer surfaces of

FIGURE 2-2. Angles cut so pieces can be joined at corners.

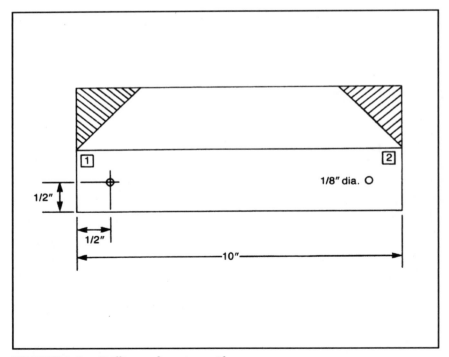

FIGURE 2-3. Drilling and cutting guide.

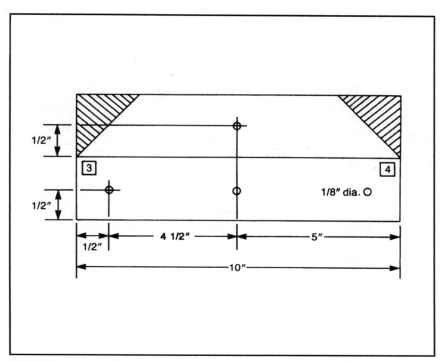

FIGURE 2-4. Drilling and cutting guide.

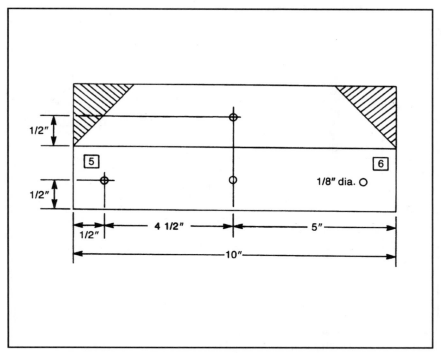

FIGURE 2-5. Drilling and cutting guide.

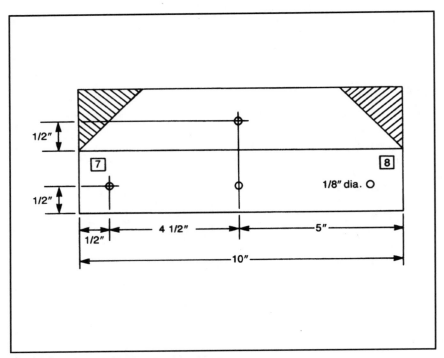

FIGURE 2-6. Drilling and cutting guide.

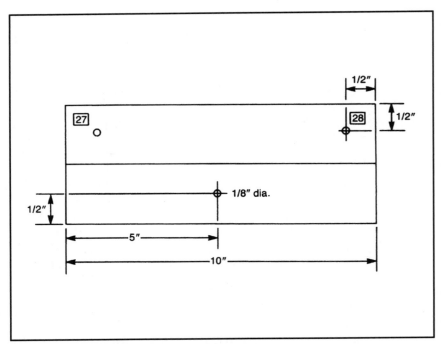

FIGURE 2-7. Drilling and cutting guide.

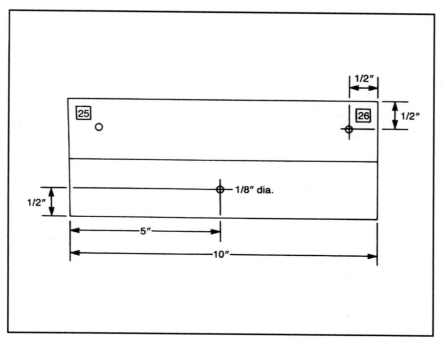

FIGURE 2-8.   Drilling and cutting guide.

that piece. The figures depict each piece as if it were laid lengthwise with its two edges on a workbench then flattened so both sides could be seen. All the figures show the locations and diameter of the holes to be drilled. The locations of the 45-degree angle cuts to be made at the ends of many of the aluminum pieces are shown as shaded areas where the aluminum is to be removed.

Most of the figures are duplicates of each other. The difference between the figures are numbers and sometimes letters that appear on the sides of each aluminum piece. The numbers are used when the framework is assembled to the two main sections that make up Questor's body. The letters are used when these sections are joined together by two connecting pieces to form the completed framework. In both cases the symbols ensure that all the predrilled holes are in their correct locations when the framework is completed. To mark the numbers and letters on the outsides of each

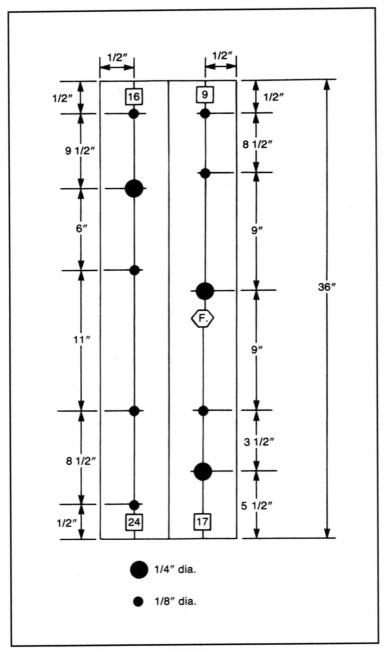

FIGURE 2-9.   Drilling and cutting guide.

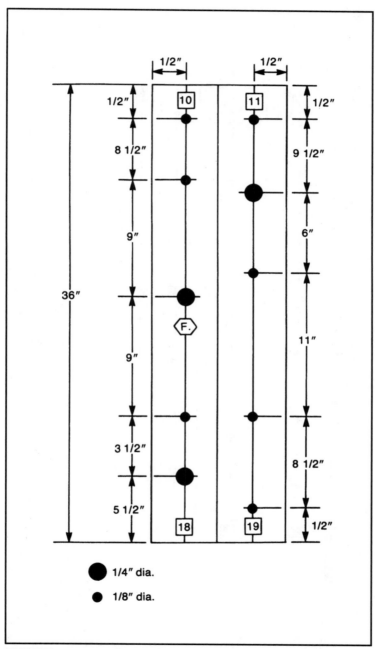

FIGURE 2-10.   Drilling and cutting guide.

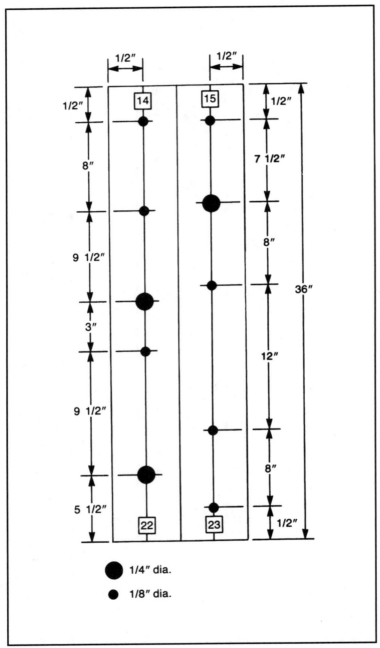

FIGURE 2-11.   Drilling and cutting guide.

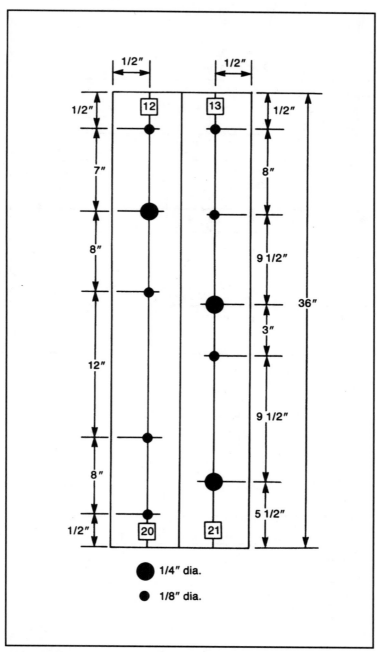

FIGURE 2-12.   Drilling and cutting guide.

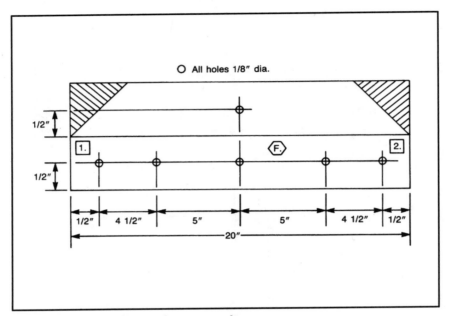

FIGURE 2-13. Drilling and cutting guide.

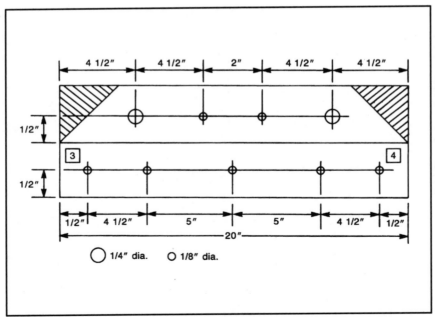

FIGURE 2-14. Drilling and cutting guide.

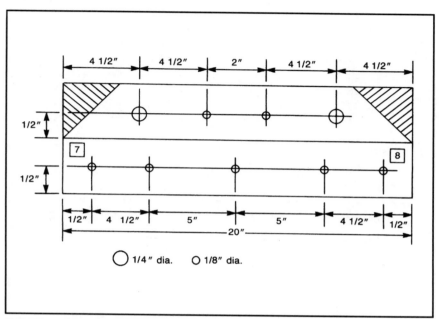

FIGURE 2-15.   Drilling and cutting guide.

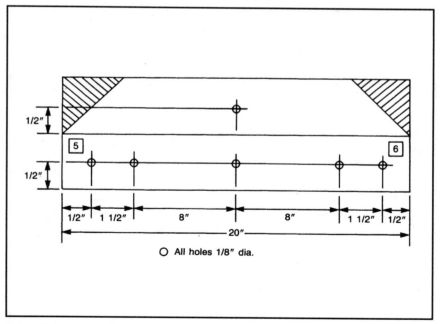

FIGURE 2-16.   Drilling and cutting guide.

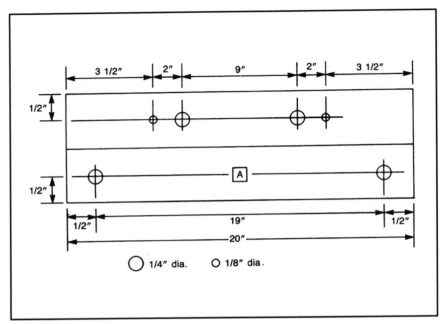

FIGURE 2-17.   Drilling and cutting guide.

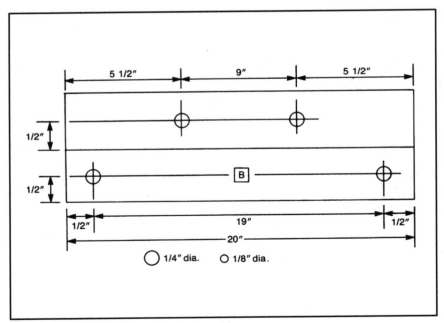

FIGURE 2-18.   Drilling and cutting guide.

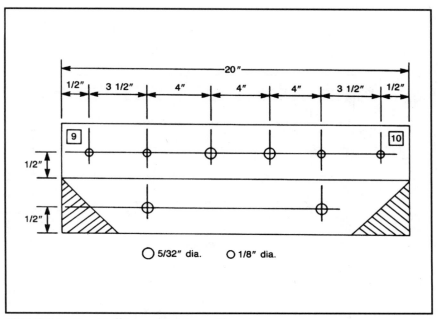

FIGURE 2-19. Drilling and cutting guide.

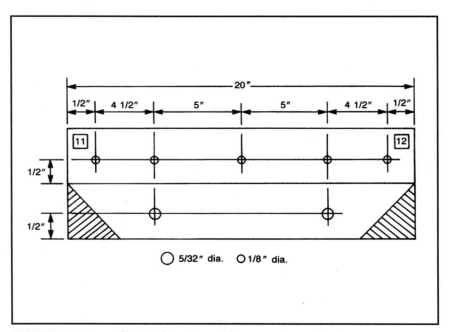

FIGURE 2-20. Drilling and cutting guide.

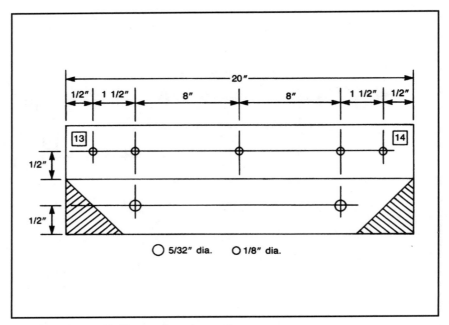

FIGURE 2-21. Drilling and cutting guide.

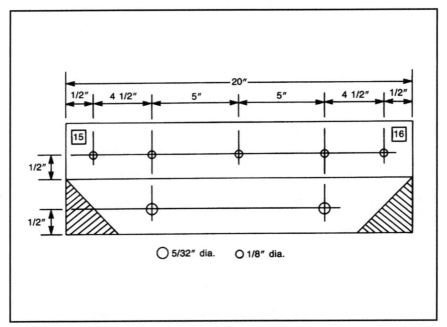

FIGURE 2-22. Drilling and cutting guide.

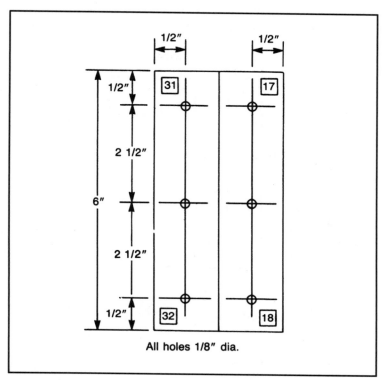

FIGURE 2-23.   Drilling and cutting guide.

aluminum section, first write the correct number or letter on a piece of masking tape and then apply the tape to its proper location according to the figure of that piece.

After all the pieces have been prepared they must again be separated into two groups—one group of pieces for each of the two sections of Questor's framework. The first group, when assembled, forms the robot's upper framework. The pieces needed are as follows: four 36-inch- and six 10-inch-long pieces.

The second group forms the lower section of the framework and the pieces needed are: ten 20-inch and four 6-inch pieces. Of the ten 20-inch pieces for the lower section of the robot, only eight are needed. The remaining two are used as connecting pieces between the lower and upper framework. These two pieces can be identified by the letters A and B which you should have marked on them.

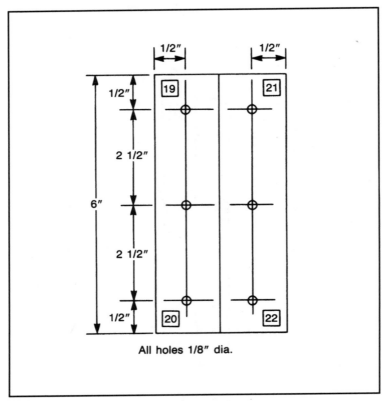

All holes 1/8" dia.

FIGURE 2-24. Drilling and cutting guide.

Now that you have the pieces regrouped, study them and familiarize yourself with these markings. How each section is assembled will become obvious as you study the pieces. The key to the assembly is the markings that allow all the predrilled holes to line up correctly.

ASSEMBLING FRAMEWORK

Assembling the framework now becomes a simple matter of matching the ends of the pieces according to the numbers on their ends; and then riveting the matched ends together. Tables 2-2 and 2-3 list the numbered ends used in the matching sequence for each of the two sections of the robot's framework. Table 2-2 is for the upper section and Table 2-3 is for the lower section.

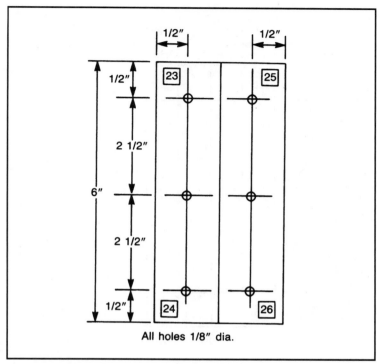

FIGURE 2-25.   Drilling and cutting guide.

After you have matched the pieces, line up the predrilled 1/8-inch holes at the ends of each piece. If the holes on the pieces do not line up (which happened to me three-quarters of the time), clamp them together and redrill the 1/8-inch-diameter hole in both pieces at the same time. Next insert a 1/8-inch rivet into the hole and using a rivet gun "pop" the rivet to secure the pieces. The rivet gun you use should come with directions.

Once you have completed both the upper and lower sections of the framework, they must be joined together with two 20-inch connection pieces marked A and B. To begin take the lower section and place it so the two Fs (which stand for front) on the front of this section face you. Then place the upper section within the lower, with its Fs facing you. Figure 2-27 shows where the two connecting pieces are placed; the piece marked A is placed at the front and the piece marked B in the rear. The pieces are held in place by four 1-inch × 1/4-inch-

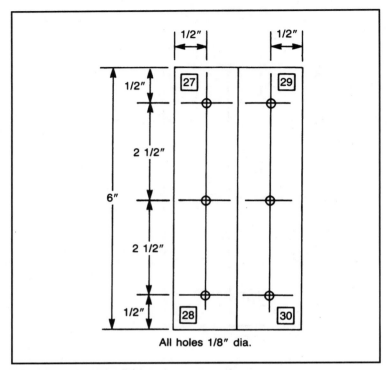

FIGURE 2-26.   Drilling and cutting guide.

TABLE 2-2. Matching List for Upper Framework

CORNERS	UPPER FRAMEWORK MATCH AND RIVET ENDS NUMBERED
1	1-9-8-16
2	2-10-3-11
3	4-12-5-13
4	6-14-7-15
5	24-26
6	23-25
7	19-27
8	20-28

NOTE: Ends numbered 17, 18, 21, 22 are not riveted.

TABLE 2-3. Matching List for Lower Framework

CORNERS	LOWER FRAMEWORK MATCH AND RIVET ENDS NUMBERED
1	1-17-8-31
2	2-19-3-21
3	4-23-5-25
4	6-27-7-29
5	9-18-16-32
6	10-20-11-22
7	12-24-13-26
8	14-28-15-30

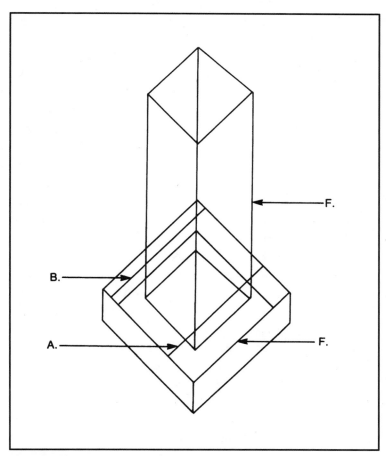

FIGURE 2-27.   Location of connecting pieces.

diameter bolt, nut, and lockwasher sets. The upper and lower body are now joined to complete Questor's framework.

MOUNTING FRAMEWORK ON THE PLATFORM

The next step in completing Questor's body is mounting the framework on the mobile platform. If you have chosen to include the vacuum cleaner option in your version of Questor, now is the time when it is installed.

Begin by placing the mobile platform on the floor so that the castor wheel faces you, this is the front of the platform. Next assemble the vacuum cleaner according to the instructions included in the kit. Place the vacuum lengthwise on the platform so that the vacuum's motor faces the rear of the platform and the hose outlet faces front. Make sure that the vacuum's mounting bracket sits flush with the platform. Once you have the vacuum cleaner positioned correctly on the platform, lower the framework onto the platform so that the Fs on the framework face the front of the platform. The vacuum cleaner should fit into the lower section of the framework as it is lowered onto the platform. You will have to move the vacuum about with the lower framework to accomplish this. Figure 2-28 shows how to position the vacuum inside the framework.

Now that you have the vacuum and framework positioned correctly on the platform, the mounting holes for each must be marked and drilled. Begin by carefully marking the two 1/4-inch-diameter holes needed to secure the vacuum cleaner unit. The bolts for this should be included in the vacuum cleaner kit. Next mark the eight 5/32-inch-diameter holes along the lower edge of the framework used to secure the framework to the platform. You will use eight 1-inch × 5/32-inch-diameter bolt, nut, and lockwasher sets to secure the framework in place. Now remove the vacuum and framework and drill the holes. Replace the framework with the vacuum cleaner inside, and bolt both in place. Press the bolts with washers in place through the bottom of the platform, then

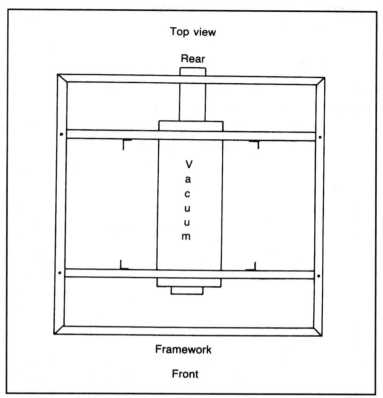

FIGURE 2-28.   Vacuum position in lower framework.

bolt the parts in place. Be sure to place lockwashers on the bolt before the nut is tightened. Figure 2-29 shows how the framework is bolted to the platform.

MOUNTING THE VACUUM OUTLET

The final step for completing the mounting procedure is to attach the vacuum cleaner outlet to the framework; this outlet consists of a hinged door and switch assembly. When the door is opened, the vacuum activates. To use the vacuum you open the door and attach a white vacuum hose to the outlet.

The outlet is held to the framework by a bracket, shown in Fig. 2-30, made from leftover aluminum angle. The outlet is

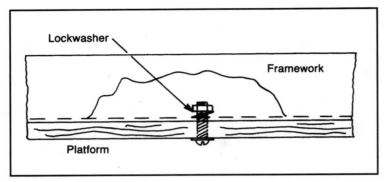

FIGURE 2-29. Bolt attaching framework to platform.

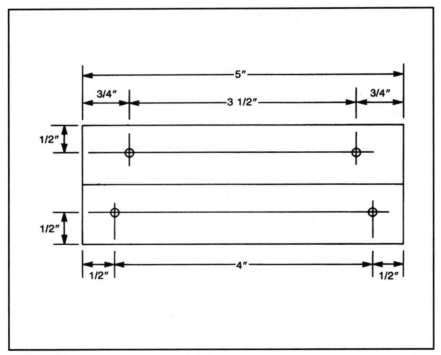

FIGURE 2-30. Vacuum outlet bracket.

FIGURE 2-31.   Mounted outlet.

then screwed to the bracket and the whole assembly bolted to the lower front section of Questor's framework utilizing two 1/4-inch predrilled holes at that location. Figure 2-31 shows where and how the outlet assembly is mounted. Now that both the vacuum and outlet are mounted, they must be joined by a short section of black hose included in your vacuum kit. Cut a piece of hose approximately 3 inches long. To connect the vacuum to the outlet, unscrew the outlet from the bracket, then connect the hose to the vacuum. Next connect the outlet to the other end of the hose and screw it back to the bracket. The mounting procedure just described is also used to change the vacuum cleaner's bags, so it is important that you understand this assembly. Questor's framework is not yet complete.

POWER SUPPLY AND TEMPORARY CONTROL BOX

Questor's systems get their power from a 12-volt battery system. This system is comprised of two 6-volt batteries. You will find as I did that most of the motors, lights, and some electronics for robots require a 12-volt system; the temporary control box is just that. Controls in the box switch the two motorized wheels on Questor's platform to on/off and reverse thus controlling his direction. There are also two speed controls mounted in the box, one for each wheel. The control box is connected to the robot's platform by a cable of wire; the length is up to you. When wiring these systems make sure you pay close attention to the wiring diagrams.

MOUNTING BATTERIES AND BARRIER STRIPS

Questor gets his power from two 6-volt, gel-type batteries mounted within his lower framework. The batteries are mounted on the right and left sides of the upper framework where it sits within the lower framework. Figure 3-1 shows where one of these batteries is located.

Each of the two batteries is held in place by three 2-inch-long pieces of aluminum angle. (Use the aluminum angle left over from the building of the framework.) Two of these pieces

FIGURE 3-1.   Location of one of the two 6-volt batteries.

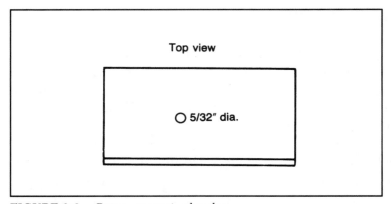

FIGURE 3-2.   Battery mounting bracket.

Table 3-1. Parts List

Amount	Item
2	6-volt, solid-gel battery, with charger kit
2	DPDT switch
1	SPST switch
1	Project box
1 or more	Roll red 18-gauge wire
1 or more	Roll black 18-gauge wire
2	2-post barrier strip and mounting screws
1	8-post barrier strip and mounting screws
2	25-watt, 10-ohm potentiometer
6	2-inch piece of leftover aluminum angle
6	1-inch × 5/32-inch-diameter bolt, nut, and lockwasher set
1	Crimp kit
4	Small electrical twist caps

are bolted to the mobile platform, while the third is bolted to the framework itself using a predrilled hole on the framework. Figure 3-2 shows where to drill holes in each of the 2-inch aluminum pieces.

To begin installing the batteries, first take two of the 2-inch pieces of aluminum angle and bolt one to each side of framework connecting piece B using two 5/32-inch bolt, nut, and lockwasher sets for each. There are two predrilled holes on each side of the upper framework. Next slide one of the batteries under each of the pieces making sure that the battery terminals face the front of the platform and that they are sitting in their correct mounting positions. Then place two more aluminum pieces, with their mounting holes flush with the robot's platform, snug against the battery. Place one piece against the front of the battery and one against the side; do this to both batteries.

Mark the mounting holes on the platform where each of the four aluminum pieces sit. Remove the pieces and battery and drill the four 5/32-inch-diameter holes in the platform

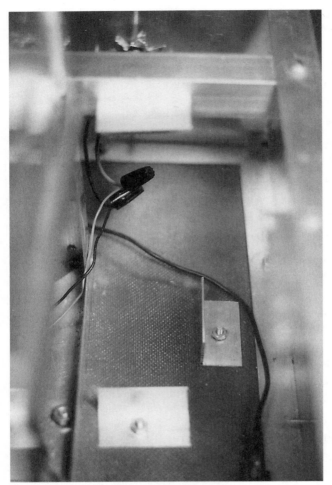

FIGURE 3-3.   Mounting brackets in place.

where marked. Once the holes are drilled, replace the batteries with terminals facing the rear, and bolt the four 2-inch aluminum pieces in place, using four 1-inch × 5/32-inch-diameter bolt, nut, and lockwasher sets. Figure 3-3 shows the mounting brackets in place. Now you could turn the robot upside down and the batteries will remain in place.

The next step in providing Questor with power is to mount multipost barrier strips at various points on the robot's platform. Figure 3-4 shows what one of these strips looks like. These barrier strips are very important because they allow the

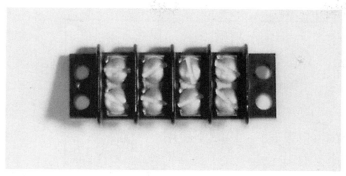

FIGURE 3-4.   Multipost barrier strip.

robot to be wired together with great ease; they also allow you to remove individual components from the robot without disturbing others. Most of Questor's electrical components use barrier strips.

For now you need only three barrier strips: two 2-post and one 8-post. These two 2-post terminals are permanently mounted on the platform near where the motorized wheel post protrudes through the platform; the exact location is of little importance. The third 8-post strip will be temporarily mounted at the center of the rear edge of the robot's platform. It will be removed later for use in the remote control system.

WIRING PLATFORM

Now that the power supply and barrier strips are mounted they must be wired together using 18-gauge wire. This wire will be used now and throughout the robot. Figure 3-5 shows a graphic representation of how the platform is wired. When you look at Fig. 3-5, you will notice that all the wires used are either red or black. The red wire represents all the wires that will eventually be connected to the positive pole of the power supply, and the black to the negative pole. While Fig. 3-5 is rather straightforward, a few things must be noted before wiring can begin.

First, the red and blue wires coming from each of the motorized wheels must be connected to their barrier strips. The wires from each wheel are too short and must be extended

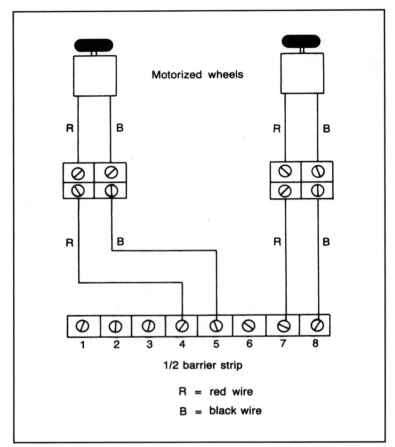

FIGURE 3-5. Platform wiring diagram.

using one 6-inch red and one 6-inch black wire for each wheel. Use twist caps or solder the red extender wire to the red wire of the motorized wheel and the black to the blue wire. To connect the extended wire to the barrier strips, twist both wires loosely together and push them up and out of the post of the motorized wheel. This post leads to the inside of the lower framework where the barrier strips are placed. Connect the wire to two of the screw posts on the same side of the strip. Refer to Fig. 3-5 for the exact connections.

Wiring the two 6-volt batteries together is made somewhat difficult because of the small size of the battery post. Instead of trying to solder the connecting wire to the battery post, I elected to use what is called a *crimp kit*. A crimp kit enables you to attach special ends to the wires that allow them to be wired together easily. Figure 3-6 shows the different ends available and the crimping tool.

As illustrated in Fig. 3-7, the batteries are not only wired together but to other components. Two of these are charging plugs that come with the batteries. Also wired between the two batteries is an SPST (single-pole, single-throw) switch. This switch serves two functions: First, it is the main on/off switch for Questor; and two, it separates the batteries when they are being charged (the switch is in the off position at this time). Make sure that you use lengths of wire long enough to allow the charging plugs and switch to reach the rear of the platform where they will be mounted later; for now you can tape the components securely to the platform.

Once you have wired the platform, use the charging plugs and charge the batteries. While the batteries are charging, it takes about 36 hours, you can construct the temporary control box used to control Questor.

TEMPORARY CONTROL BOX

Before you begin to assemble the temporary control box, a brief explanation of how it functions is in order. To begin, the two 6-volt batteries have been wired together to give Questor a 12-volt power source. This power source is then wired to two potentiometers, one for each motorized wheel, within the control box. These *pots* as they are commonly called, are a type of variable resistor that lowers or raises the voltage coming from the batteries. The pots are used to control the speed of each motorized wheel.

The lowered or raised voltage passes into two double-pole, double-throw (DPDT) switches, again one switch for each motorized wheel. The DPDT switches are actually two switches in one, hence the term double in their description. To reverse the direction of a dc electric motor, you must

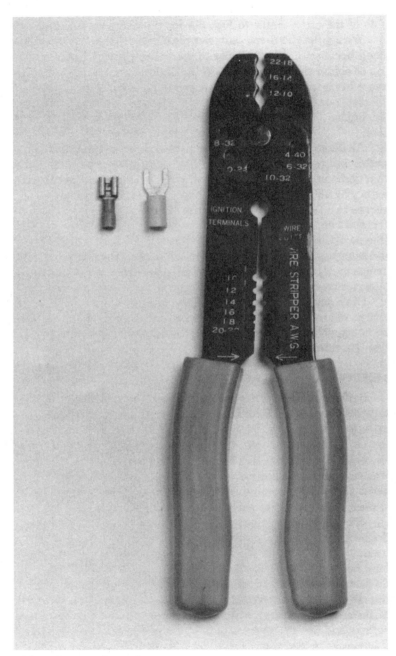

FIGURE 3-6.   Crimp kit.

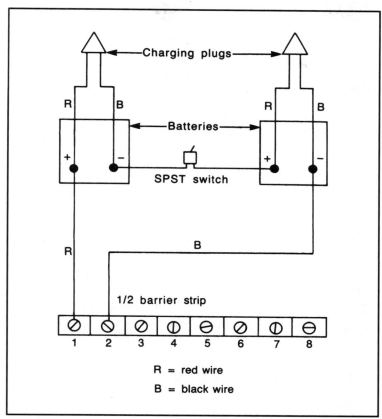

FIGURE 3-7.   Battery wiring diagram.

change the polarity of the wires leading to the motor. For example, if the right terminal of the motor is connected to the positive terminal of the power source, and the left to the negative, the motor will run clockwise. Exchange the leads so the right lead is negative and the left positive and the motor will run counterclockwise, or in reverse. The DPDT switch does all this internally so all you do is flip the switch up or down to change the direction of the motor. Also included in these switches is a center on/off position where no power goes to the motor.

After passing through the DPDT switch the voltage reaches one of the two motorized wheels on the robot's platform, and depending on the position of the switch the motor will run forward, reverse, or not at all. How this system is used to control Questor will be described later in the chapter.

CONTROL BOX CONSTRUCTION

The temporary control box will house all of Questor's control electronics in this stage of his construction. The box itself should be approximately 4 × 4 inches square to allow room for the various parts. The parts contained in the control box are two heavy duty DPDT switches and two potentiometers like those shown in Fig. 3-8. These components are wired together in the control box then connected to the robot's batteries and motorized wheels via a group of wires taped together in a cable. How the parts are mounted in the control box is up to you; however, Fig. 3-9 shows a recommended layout. To mount the parts you will have to remove the box's overplate on the control box and drill mounting holes in that plate.

WIRING THE TEMPORARY CONTROL BOX

The wire used in the temporary control box and throughout the robot is an 18-gauge-type colored either black or red. Again, red is for all wires connected to the positive pole of the batteries and black is for all to the negative. This makes it easier to trace

FIGURE 3-8. DPDT and "pots" switch.

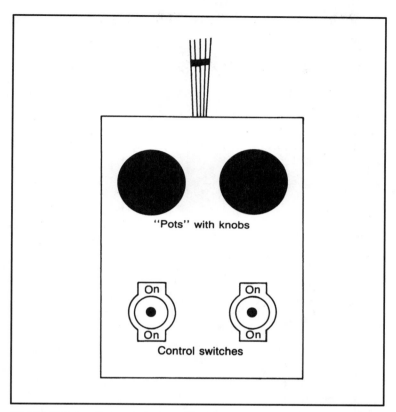

FIGURE 3-9.   Suggested control box layout.

the various circuits in Questor. The robot's electronics are not so complicated that you would confuse these wires with others leading to Questor's various systems.

Figure 3-10 shows how to wire the components in the control box together. The color of each wire has been noted. Try as I might, I was unable to make Questor a completely solderless project. You will have to solder some of the robot's components. Two of these components are the pots in the control box. If you have never soldered before, you could simply twist the wires around the post of the components, but this makes for loose and many times poor electrical contacts. What you can do is twist the wires now and solder them later when you have picked up the skill.

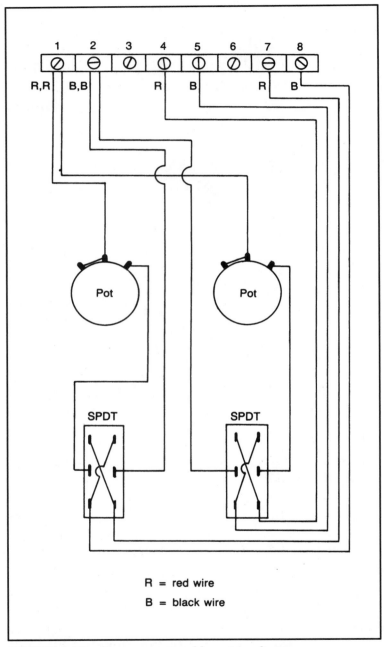

FIGURE 3-10.   Temporary control box wiring diagram.

When you look at Fig. 3-10, you will notice eight numbered wires coming out of the control box to the barrier strip on the platform. The numbers on each wire correspond with numbers on the posts of the 8-post barrier strips located at the rear of the robot's platform. Match the numbers on the temporary control box to the platform to complete this phase of Questor's construction.

USING THE CONTROL BOX

The temporary control box is very simple to use. The first thing to do is to activate Questor by flipping the main power switch to the on position. Turn both pots on the control box all the way to the right and then turn the pots slightly to the left. This reduces the speed of both wheels so the robot will travel slow enough for you to get familiar with using the control box. Next flip the two DPDT switches on the control box up and Questor will begin to move slowly forward. If you flip them down he will move backwards. Allow the robot to travel for about 20 feet. You may notice that Questor is veering either left or right. You can correct this using the two pots on the control box. If the robot is veering left, increase the speed of the left wheel slightly, and if he is veering right, increase the speed of the right wheel. This should straighten out Questor while keeping his speed up. You could also straighten the robot's path by decreasing the speed of the wheel opposite the direction he's veering. This, however, also slows the robot down and if you are already operating Questor at a slow speed, this could slow him down too much. Later you can use the pots to increase Questor's speed and then recalibrate his direction.

Turning the robot can be accomplished in one of three ways. The first is to run one motorized wheel forward and the other in reverse; this allows Questor to turn about his center. This method comes in very handy when operating the robot in close quarters. The second way to turn Questor is to turn one wheel off and run the other either forward or reverse, depending on the direction in which you want to go. By steering the robot in this way, you can make his turns wider and smoother looking. The final method of directing Questor only works with the temporary control box. The pots on the control box are used to vary the speed of the motorized wheels allowing one to overpower the other and veer the robot in the desired direction. This of course is the opposite of what you did to straighten Questor's path. The remote control system does not have a speed control function built into it, so this method of control cannot be used with this system; but that is not to say you could not design this capability in your robot.

At this point you have completed most of the major work in Questor's construction. Now is the time to experiment with the robot's control and refamiliarize yourself with the rest of the book. The next chapter details Questor's remote control system. If you do not plan to include a remote control system or may plan to include one later, you may skip that chapter. I would, however, recommend you do read it to give you an understanding of remote control systems.

REMOTE CONTROL SYSTEM

Wireless control has always seemed to fascinate people, and Questor's remote control system is the heart of his appeal. While the technical aspects of remote control may be a little hard for the novice to grasp, Questor's remote control system is rather simple in construction. Before I go into detail on how the system is comprised, a brief explanation of remote control is in order.

A remote control system consists of three basic components. The first is the transmitter or "encoder." Moving controls on the transmitter causes it to send or encode signals to the second part of the remote control system, the receiver, or decoder. The receiver gets the signals from the transmitter and then decodes them. Depending on what signal the receiver decoded, it will activate a servo, the third part of the system. Servos are the mechanical part of a remote control system. A wheel or sometimes bar on the servo will turn in proportion with the movement of the transmitter's control. This movement can then be used to directly control the function of a robot, or in Questor's case to trip switches that control his movements.

Questor's remote control system is a standard off-the-shelf type like that pictured in Fig. 4-1. Notice the three main parts of the system. The robot requires a system with a minimum of two channels. A two-channel system has two servos; each of the servos is used to control one of the robot's motorized wheels. The system used in my version of Questor has three channels; the third channel is used to trip two switches that can turn other items on the robot on or off.

FIGURE 4-1.   Three-channel remote control system.

FIGURE 4-2.   Leaf switch.

The switches that the servos trip are called leaf switches (Fig. 4-2). A leaf switch is a very small on/off switch that is triggered by depressing a small metal strip or "leaf" on the switch. By using four leaf switches, it is possible to recreate the function of the DPDT switches used in the temporary control box.

A total of eight switches is needed to duplicate the function of the DPDT switches used to control the robot's motorized wheels. One servo is then used to trip four switches in such a way to drive the wheel either forward or reverse. You use the control sticks on the remote control transmitter in the same way as you flipped the DPDT switches on the temporary control box; up is forward, center is off, and down is reverse. If you chose a remote control system with more than two channels, you can use the other servos to trip leaf switches for turning other devices on or off, or control motors (forward, stop, and reverse) within the robot. The third servo of my remote control system is used to turn a horn on and off.

TABLE 4-1. Parts List

Amount	Item
1	2-to-3-channel remote control system
10	Leaf switch and mounting screws
1	10- × 10- × 1/8-inch plywood square
4	1- × 10- × 1/8-inch wood strip
4	8-post barrier strip
4	Screw hook
2	Rubber band
1	Small strip of foam rubber
#	Spools of 18-gauge black-and-red wire
4	2- × 2-inch aluminum corner brace
4	2-inch × 1/8-inch-diameter bolt, nut, and lockwasher set
1	4-slot fuse holder
4	SFE 20-amp fuse

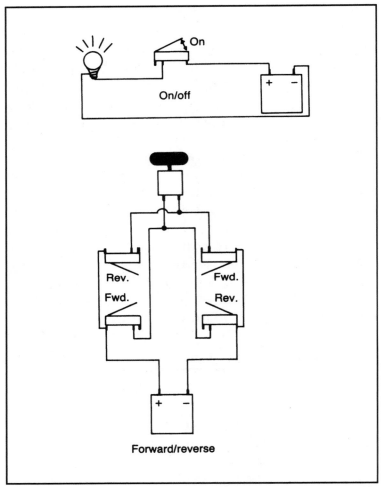

FIGURE 4-3. Control options using leaf switches.

You need only one leaf switch per function if that function is to be turned only on or off. Figure 4-3 shows how the leaf switches are positioned and triggered for either on/off or forward/reverse control. By now you're probably wondering where all this fits inside of Questor. The remote control system (servos and receiver), leaf switches, and other components are mounted on a motherboard that is then installed inside Questor's framework.

MOTHERBOARD

The motherboard is simply a 10- × 10- × 1/8-inch piece of plywood on which all of the components for the remote control system are mounted. The various components consist of the remote control system's servos, receiver, and battery pack, along with ten leaf switches, four barrier strips, and a four-slot fuse holder. Figure 4-4 shows where each item is placed on the board. The first items to be mounted are the servos.

Cutouts will have to be made in the board to allow the servos to sit flush with the board. To do this, first place the servos evenly spaced on the motherboard and trace around their bases. Cut out the wood where traced and slip the servos in place. The servos' body should have tabs sticking out along its top edge; these tabs prevent the servo from going all the way through the board and this is where the servos are screwed to the board. Most remote control systems come with either plastic wheels and/or star levers that are screwed on the servo's

FIGURE 4-4. Parts layout of motherboard.

FIGURE 4-5.   Star lever.

motor. Figure 4-5 shows how a star lever looks. For this application you will need star levers that must be modified by removing four of the star's legs. What you end up with is a straight bar like that in Fig. 4-6. The bar will trip a bank of leaf switches on either side of the servo.

FIGURE 4-6.   Modified star lever.

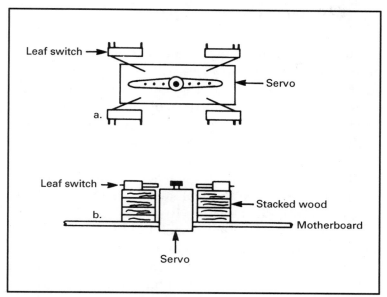

FIGURE 4-7.   Leaf switch mounting.

In order to be tripped, the leaf switches must line up with the trip bar on the servo. This is accomplished by stacking 10- × 1- × 1/8-inch strips of wood along each side of the servos, then mounting the leaf switches on the strip. Figure 4-7 shows where to mount the leaf switches on the wooden strip in relation to the servos. Once the leaf switches are mounted, connect the servos to the remote control system's receiver and by using the transmitter, check to see if the trip bar is activating the switches properly. The order in which the servos are wired to the receiver is not important at this time; however, when the receiver is mounted, the sequence will be detailed so that the control stick on the transmitter activates the proper servo.

Once the servos and leaf switches are operating properly, the barrier strips and fuse box for the remote control system can be mounted to the motherboard. Figure 4-4 shows where to mount these components. With all components mounted, the next step is to wire them together.

WIRING THE MOTHERBOARD

Before you wire the motherboard, cut two notches on each side of the motherboard so the wires will not go past its edge. Figure 4-8 shows how to wire together the components on the motherboard. There are two main rows of barrier strips on the motherboard; the first row is numbered. These numbers correspond with numbers on the tabs of the leaf switches; simply wire the matching numbers together. In some cases more than one wire will go to one post on the barrier strip. Use the half of the barrier strip closest to the leaf switches. The color wire used is indicated on the leaf switch: R = red, B = black. The other row is where the motorized wheel and horn will be connected; they too use the matching number system.

The second row is divided into two parts called power grids. The first 8 post (which is one complete barrier strip) is called the positive grid and is where the positive lead of the battery is connected and where all the positive or red wires from Questor's electronics will be connected. The second 8 post is for the negative or black wires and is called the negative power grid. All the posts on the same side of each grid must be wired together by one wire run from post to post. Be sure not to run a wire between the positive and negative grids; this will cause a short circuit. Figure 4-8 shows where the wire runs. Later when other functions are wired, the instructions will say "wire to positive power grid and negative power grid." You can then connect those wires to any open post on the grids. Figure 4-8 also shows four wires coming from the positive grid to the fuse holder. These wires are all positive and you should use red wires. Two more red wires run from the opposite ends of two of the fuses directly to the post on the leaf switch barrier strips. This is where the switch gets the power to control two on/off functions in the robot. (The negative or black wire forms the function being controlled; in my robot a horn is wired directly to the positive power grid.) There are also two black wires running from the negative power grid to the leaf switch barrier strips at post 8 and 2. These are also shown in Fig. 4-8.

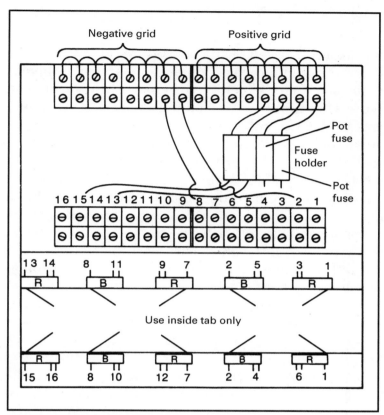

FIGURE 4-8. Motherboard wiring diagram.

Wires to the leaf switches and fuse holder will have to be soldered. The wires that lead to the barrier strips should have hooks bent at their ends so they can wrap around the screws on the strip. After the board is wired, check it against Fig. 4-8 because errors here can affect the function of the rest of the robot. Also at this time, install four 20-amp fuses in the fuse holder. These fuses help protect the robot's components from short circuits and overloads. Once the board is wired and checked, the remote control receiver can be mounted and the motherboard mounted in Questor's framework.

COMPLETING THE MOTHERBOARD

The remote control's receiver and battery are mounted on the underside of the motherboard. Using four screw-on hooks, rubber bands and foam rubber, the receiver is held securely in place. Figure 4-9 shows how to mount the receiver. The figure is self-explanatory. The only thing to keep in mind is that the servos must be wired to the receiver, so don't mount the receiver out of reach of the servo wires.

The order in which the servos are connected to the receiver is very important to the control of the robot. When both control sticks on the transmitter are pushed up, the robot should move forward. If both sticks are pulled down, the robot should run in reverse. The center or neutral position is off and of course causes no movement of the robot. If you have a third channel (and servo) in your remote control system, it should react to the sideways movement of one of the control sticks on the transmitter. Table 4-2 lists all of the control combinations used to operate Questor's functions.

It is not necessary to wire the motorized wheels to the motherboard. To check this simply make sure that when the sticks are pushed forward, the two servos controlling the motorized wheels turn as shown in Fig. 4-10. If you have a third servo a sideways movement of either stick should cause the servo to activate it.

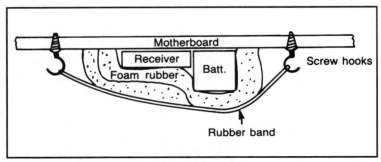

FIGURE 4-9.   Receiver and battery mounting.

TABLE **4-2.** Control Combinations Using Transmitter Control Sticks

STICK POSITION		ROBOT MOVEMENT
Right	Left	
Up	Up	Forward
Down	Down	Reverse
Center	Center	Stop
Up	Down	Circle right
Down	Up	Circle left
Up	Center	Turn right
Center	Up	Turn left

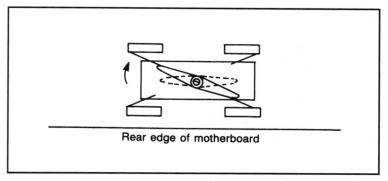

Rear edge of motherboard

FIGURE 4-10.   Proper servo activation to trip leaf switches.

INSTALLING AND WIRING THE MOTHERBOARD

After the servos have been checked, the motherboard can be installed in Questor's framework and wired in place. To mount the board you will need four 2- × 2-inch aluminum corner braces available at any hardware store. These are bolted in place at the lower part of the robot's upper framework where the four bolts holding the two connecting pieces of the framework are located. Figure 4-11 shows one angle in place. The motherboard is then attached to these angles.

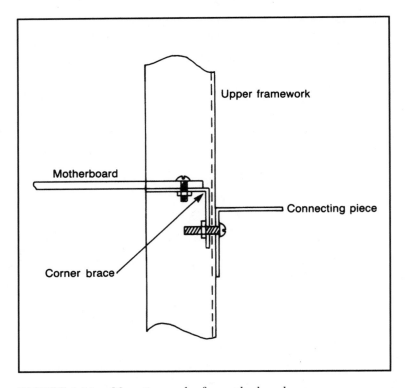

FIGURE 4-11. Mounting angles for motherboard.

Mark the holes for the aluminum angles on the mother-board from underneath. Then remove the motherboard and drill four 1/8-inch-diameter holes where marked, being careful not to damage the parts already mounted on the motherboard. Attach the motherboard to the angles using four 2-inch × 1/8-inch-diameter bolt, nut, and lockwasher sets.

Wiring the motherboard to the rest of the robot now becomes a simple matter of matching number wires from the motorized wheels and battery system to numbered posts on the barrier strips. The two pots used to control the robot's speed are wired in at this time. One wire to each pot comes directly from the fuse holder. Figure 4-12 shows the connection to be made for the entire system. Since Questor has yet to get his metal skin, the pots have no place to be

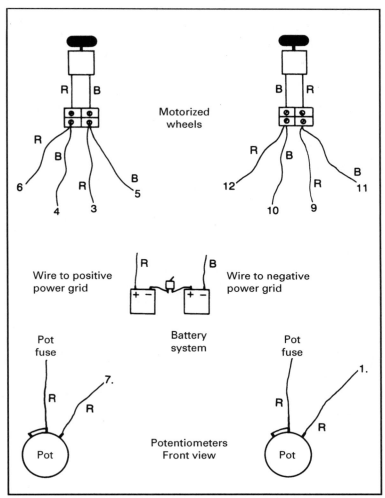

FIGURE 4-12.   Platform-to-motherboard wiring guide.

mounted on the robot, so for now simply tape them on the platform at the rear of the robot where they will later be mounted. Many of the spaces on the power grid barrier strips will remain empty until later in the robot's construction. As each new function is added to the robot, the empty spaces will be used.

USING THE REMOTE CONTROL SYSTEM

The remote control system operates much in the same way as the temporary control box did. In the control box the position of the switches, up for forward, center for off, and down for reverse, along with their combinations, dictated the direction of the robot. With the remote control system, the control sticks on the transmitter take the place of the switches of the temporary control box. Unlike the temporary control box, however, the transmitter cannot control the speed of the robot because the pots are no longer at the controls but on the robot itself. This means that you will have to calibrate Questor's motorized wheels and preset the robot's speed before you use the remote control system, a small price to pay for wireless control. As I noted before, if your remote control system has a third channel and thus a third servo, you could remotely control other robot functions. This extra servo is controlled by the sideways movement of one of the control sticks.

Congratulations! You have just completed the last major component of Questor's construction! The rest of the book covers various subsystems within Questor as well as the cosmetic aspects of the robot's construction. If you plan on building a duplicate of Questor, the following chapters detail the rest of his construction. However, it is my hope that you will use them as a guide to create your own individual robot servant.

ARMS AND SUBSYSTEMS

In this chapter you will fabricate and assemble Questor's arms, drink dispenser, and head as shown in Fig. 5-1. Also you will wire the vacuum cleaner completing that system. Questor's arms and drink dispenser will be built first because they are interrelated to one another. The pump and tank for the drink dispenser are housed inside the robot's body while the fluid outlet and control button are mounted on Questor's arm—more specifically, his left wrist. The drink dispenser operates in a rather straightforward manner; pushing the control button on Questor's wrist activates a small (and slightly noisy) 12-volt pump. This pump draws fluid out of a 1-gallon

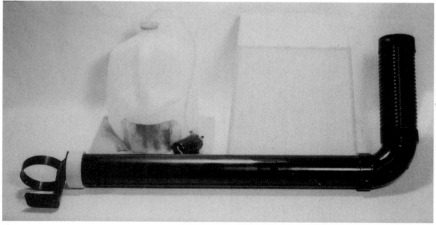

FIGURE 5-1. Arm, drink dispenser, and head.

container (a former milk container) and passes it via a tube (fish tank air tubing) to an outlet on the underside of the robot's wrist below the control button. This way the person getting the drink can control the amount dispensed.

Questor's head consists of a 12-volt automobile dome light and a cube shaped cover made of opaque ceiling light panels. The head's cover is the most delicate part of Questor and should be fabricated, fitted, and then removed until the

TABLE 5-1. Parts List

AMOUNT	ITEM
1	Pushbutton SPST switch
1	12-volt dc pump
1	1-gallon milk container and cap
1	8-foot × 2-inch PVC piping
4	2-inch-diameter PVC end-cap
2	2-inch-diameter PVC elbow
1	9 1/2- × 9 1/2- × 1/8-inch sheet of hardboard
4	1-inch × 1/4-inch-diameter nut, bolt, and lockwasher set
1	1- × 2-foot section of sheet metal
6	1/4-inch-diameter washer
1	4-foot-length fish tank air tubing
2	Auto drink holder
1	Tube of silicone glue
1	1-slot fuse holder
1	12-volt auto dome light with switch
1	2- × 4-foot sheet of ceiling light panel
6	1/8-inch × 1/8-inch-diameter sheet-metal screw
#	Miscellaneous wood screws
1	2-post barrier strip
#	Rolls of 18-gauge wire, red and black
4	2- × 2-inch corner brace
4	1-inch × 1/8-inch-diamter bolt, nut, and lockwasher set

robot's final assembly to protect it from damage. The head lights when a button (that should come with the auto dome light kit) mounted on the robot is pushed. Questor's head serves two functions: One, as an attention getter, being that the light is quite bright and, two, to illuminate the objects being carried on his serving tray, The serving tray will be mounted at the same time as Questor's head, during final assembly. The following sections will detail the construction and installation of the robot's arms, drink dispenser, and head.

ARMS

Questor's arms are very simple to make and install; however, care should be taken when fabricating them because errors in the parts may affect the way the arms are mounted and how level the serving tray sits on the arms. After viewing pictures of Questor earlier in this book, the way the arms look and their position should be rather obvious.

The arms are made from 2-inch-diameter PVC piping used for plumbing (or robot arms). You will need two 8-inch and two 21-inch lengths of PVC. When you buy the PVC, also purchase two 90-degree turn elbows and four end-caps. Figure 5-2 shows what these parts look like.

When you cut the PVC piping, be sure to make your cuts as straight as possible; I suggest you use a miter box to ensure straight cuts. Cut the lengths of PVC according to Fig. 5-3 and assemble them, as well as the elbows and end-caps, together according to the figure. Be sure that the elbows and end-caps fit snugly on the tubing because you will not be gluing or fastening the parts together.

Once assembled, check to see that the arms are the same height and length. Do this by standing them next to each other. If the arms are not correct, all that may be needed is some tapping or pulling of the end-caps so that they fit more snugly or closer to the PVC pipe, or stick out farther from it. If the deviation is large, you should disassemble the arm and recut a new section of PVC piping. The piping is usually sold in eight-foot

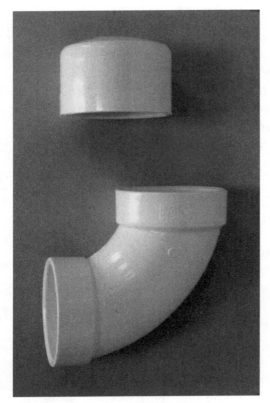

FIGURE 5-2.   Elbow and end-cap.

lengths, so you should have plenty if this occurs. After you have the arms assembled, take a marker and draw a ring around the pipes tracing the ends of the elbows and end-caps. These lines will act as a guide, so if you disassemble the arms for cutting and drilling, they can be reassembled correctly.

Now take one of the arms and hold it up against the side of Questor's upper framework so that the top edge and lower arm lines up with the two 1/4-inch-diameter predrilled holes on the framework as shown in Fig. 5-4. Mark where the holes meet the elbow and arm and drill two 1/4-inch-diameter holes where marked. Do the same with the other arm. Before the arms can be bolted to the framework, a 3/8-inch hole must be drilled on the outside of the lower part of the arm. Figure 5-5

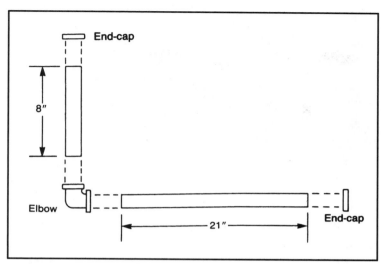

FIGURE 5-3.   Arm assembly.

FIGURE 5-4.   Mounting holes for the arm.

FIGURE 5-5.   A 3/8-inch-diameter access hole.

shows the location of this hole. The hole which lines up to the 1/4-inch-diameter hole drilled previously is used to access the head of the bolt so you can turn it with a screwdriver when attaching the arm to the framework.

Three more holes must now be drilled in Questor's left arm to accommodate parts used in the drink dispenser. One of the two 1/4-inch-diameter holes is drilled along the inside of the lower section of the arm and allows a tube from the drink dispenser to pass into the arm. The second 1/4-inch-diameter hole is drilled on the underside of the arm just behind the front end-cap. This hole is where the outlet spout of the dispenser is located. The final hole drilled in the arm is located directly above the hole for the outlet spout on the top of the arm behind the end-cap; it allows wires for the control button that switches the drink dispenser on and off to reach the switch that will be mounted in the end-cap. The hole's diameter depends on the type of switch you have obtained. Figure 5-6 shows where the three holes are drilled on the left arm.

After all the holes have been drilled, there is one final preparation to be made. Two small squares must be cut from the ends of the two upper pipes of the arms. These cutouts allow for the pipes to be refitted to the elbows once the bolts are in place. Make sure these cutouts are bigger than the head of the bolts you use. Refer to Fig. 5-7 for the location.

The arms can now be reassembled and bolted to the framework. Starting with the lower portion of the arm, reattach it to the elbow, then aligning the hole on the elbow once

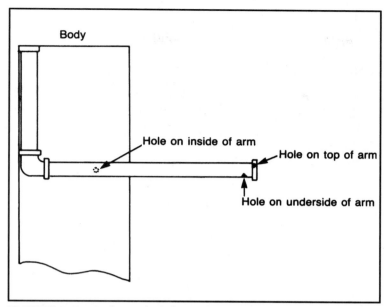

FIGURE 5-6. Holes in the left arm for the drink dispenser.

FIGURE 5-7. Bolt head clearance cutouts.

again with that on the frame, bolt it in place using a 1-inch ×
1/4-inch-diameter bolt, nut, and lockwasher set. Do the same
with the mounting hole on the lower arm, only stack three
washers between the arm and the framework so the arm sits
straight along the framework. You can access the head of the
screw through the 3/8-inch holes drilled on the sides of the
arm. Now you can replace the upper section of the arm being
sure to align the small cutout section on the piece with the
head of the bolt holding the elbow to the framework. Figure 5-8
shows how the arms are mounted to the framework.

The last component that completes Questor's arms are his
hands. The hands are simply two auto drink holders mounted
to the end-caps on the front of Questor's arms. The top of the
drink holder is removed and the rest is screwed to the end-cap
with a 1/8-inch screw. Figure 5-9 shows a completed hand.
After you have made the hands, set them aside; they will be
attached to the robot's arms later during the robot's final
assembly. You are now ready to build Questor's drink dispenser
and install it in the robot's left hand.

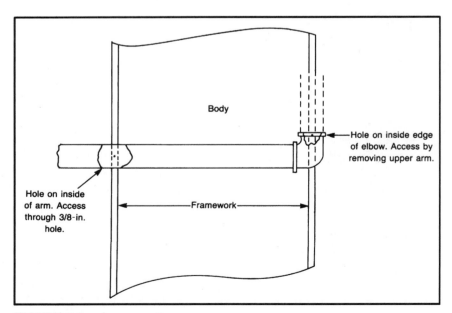

FIGURE 5-8.   Arm mounting.

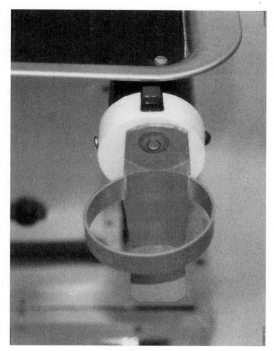

FIGURE 5-9. Completed hand.

DRINK DISPENSER

Figure 5-10 shows the parts layout for Questor's drink dispenser. The dispenser's base, on which all of the components are mounted, consists of a 9 1/2- × 9 1/2-inch piece of 1/8-inch plywood. The unit, when completed, is mounted in Questor's upper framework, hence the 9 1/2- × 9 1/2-inch base. As you can see by the previous figure, many parts for the dispenser must be modified or fabricated before they are mounted; the first of these is the 12-volt pump itself.

The pump (listed in parts list; your pump could be different) has two large input/output spouts that have to be removed so that the remaining parts of the spout can accept the narrower tubing used to pass fluids through the system. Figure 5-11 shows what portion of the spout to remove. Next, two sheet metal brackets need to be cut; these brackets hold the drink dispenser's tank, a 1-gallon milk container, in place and prevent it

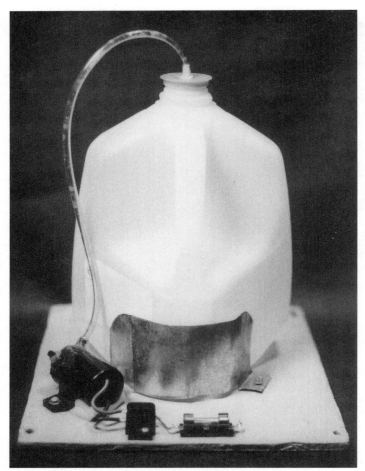

FIGURE 5-10.   Drink dispenser layout.

from shifting when the robot moves. The template for those brackets is shown in Fig. 5-12. Lastly, the cap to the 1-gallon milk container must have two holes made in it. The first hole, located in the center of the cap, is for the input tube going to the pump that draws fluid out of the container. A second smaller hole is made next to that one so air can get into the container when fluid is being drawn out. Otherwise, a vacuum would form inside the container, causing it to collapse.

After the parts are prepared, they can be mounted on the base as shown in Fig. 5-10. Start with the brackets that hold

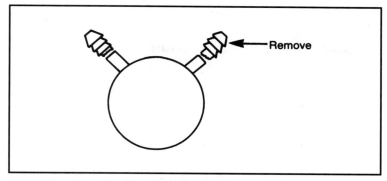

FIGURE 5-11.   Remove these sections from spout.

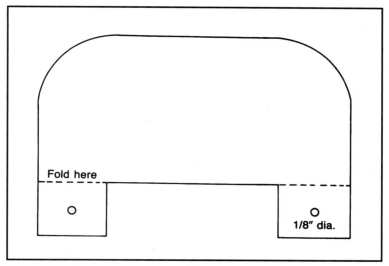

FIGURE 5-12.   Template for drink tank brackets.

the 1-gallon milk container in position. Place the container on the base and press the metal brackets around it; the metal will bend to the shape of the container. Now mark the location of the brackets mounting tabs, remove the container, and screw the brackets in place using four 1/8-inch wood screws.

Next is the pump which is mounted in the right rear corner of the base. It too is held in place by 1/8-inch wood screws. Near the pump on the back edge of the base, mount the barrier strip and fuse holder; make sure to allow about an inch between each component. The last thing to do at this stage is to connect the pump to the container using a 2-foot piece of fish tank air hose tubing; the tube should go from the pump through the cap of the container and rest on the bottom of the container. Smear some silicone glue around the connection at the pump to ensure a waterproof fit. The dispenser is now ready for mounting and wiring.

The drink dispenser is held in place using four 1-inch angles like those used in the remote control system. There are four predrilled holes in the framework to accept the angles that are held in place by four 1-inch × 1/4-inch-diameter bolt, nut, and lockwasher sets. Bolt the angles in place, then position the drink dispenser on them. Take a pencil or pen and mark via the underside of the base where the unit sits on the angles. Remove the drink dispenser's container and drill four 1/8-inch-diameter holes where marked; be careful not to damage the parts already mounted on the board. Now bolt the unit in place using four 1-inch × 1/8-inch-diameter bolt, nut, and lockwasher sets. Replace the container on the base and reconnect the tube from the pump to the container. Once the drink dispenser is in place, a second length of tubing must be run from the output spout of the pump into the robot's left arm and out of a predrilled hole at the bottom of the arm. The tubing should be approximately 20 inches in length. Also smear some silicone glue where the tubing connects to the pump. Figure 5-13 shows how to run the tubing through the arm and how to form the spout at the robot's wrist where people fill their cups. The last step in installing the drink dispenser is to wire the pump, control switch, and power together. Figure 5-14 shows how to wire the dispenser in place. Note the use of a 20-amp fuse; this fuse is short circuit protection for the pump, and the wire leading to its holder must be soldered in place. The control button mounted on the robot's wrist will also have to be soldered with two wires running the length of the arm, one that runs down the framework to the barrier strips on the remote control board.

After the dispenser is wired in place, it must be tested. Place a small amount of water in the container, then turn on

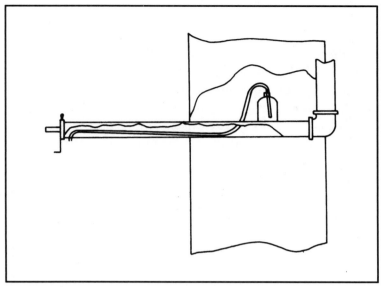

FIGURE 5-13.   Tank-to-arm tubing.

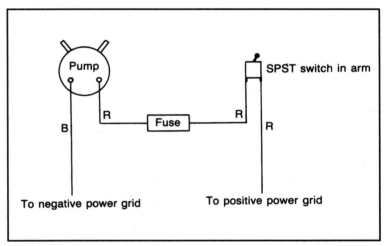

FIGURE 5-14.   Drink dispenser wiring diagram.

the robot's main power switch and press the switch on the robot's wrist. The pump will begin to hum and once it primes itself, water will flow out the spout at the underside of the robot's wrist; this completes the drink dispenser. Next you'll install Questor's head.

THE HEAD

Questor's head is basically an automobile dome light with a white cover made of ceiling light panels. The dome light is attached to a 9 1/2-inch-square piece of 1/8-inch pressboard (or plywood; hardboard is easier to cut circles in) which is in turn mounted at the top of Questor's upper framework using three predrilled holes located on the framework. Three metal tabs located where the light's base is screwed to the framework act as mounting points for the light's cover. Figure 5-15 shows the dome light and cover mounting tabs attached to the robot's frame.

The mounting tabs for the cover are made from the same type of sheet metal used to make the container brackets for the drink dispenser. The tabs should be 2 inches long folded in the middle with a 1/8-inch hole drilled at both ends. Note that the holes drilled in the ends of the tabs should be slightly smaller in size than the diameter of the 1/8-inch sheet metal screws used to hold the dome light assembly to the robot's frame.

The dome light cover, which is actually considered Questor's head, is made from a sheet of plastic ceiling light

FIGURE 5-15.   Mounting tabs for the head. Note the auto dome light in place.

panels and eight lengths of plastic angle. The ceiling light panels are available at most hardware stores and the plastic angle you need can be purchased at a hobby shop. Figure 5-16 shows the size, shape, and amount of the panels to be cut from the sheet of plastic ceiling light cover. When cutting the plastic panels, use a plastic cutting knife that should be available where you obtain the panels.

Assemble the panels as shown in Fig. 5-17 using the lengths of plastic angle where the panels meet, and modeling glue to cement the parts together. After the glue has dried, place the cover over the dome light so that the tabs on the framework are inside the cover. The cover is clear enough so that you will be able to see and mark the hole in the tabs on the cover. Carefully drill the mounting holes in the cover making sure not to crack the plastic. Now attach the cover with three 1/8-inch sheet metal screws; you can now wire the dome light.

Figure 5-18 shows how to wire the dome light. The switch (which should come with the dome light) for the light will later be mounted on the robot's skin; for now, simply tape it to the robot's frame. Test the light being sure to flip Questor's main power switch on before trying the light. Now that the head has been wired and tested, remove the cover so it will not be damaged while you complete Questor's construction.

WIRING THE VACUUM SYSTEM

The final major system to be wired is Questor's vacuum system, the body and motor of which you installed earlier. The wiring guide that comes with the kit is for installation of the vacuum in a car, while basically the same type of installation as here, a few things should be noted. First, the kit comes with all black wires, not black and red, and the motor itself has two white leads. The motor was designed to run in the right direction no matter what lead is connected to positive or negative terminals of a battery. Just be careful not to connect both leads to the same terminal or in this case, the power grid. Also, on the switch you will see three tabs where the wires are

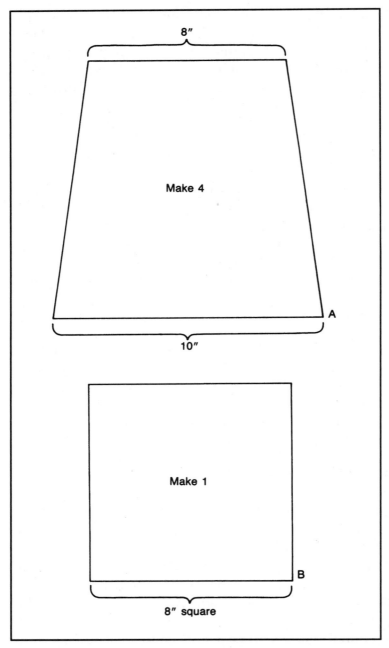

8″

Make 4

A

10″

Make 1

B

8″ square

FIGURE 5-16.   Head panels to be cut from ceiling light panel.

FIGURE 5-17.   Inside view of assembled head panels.

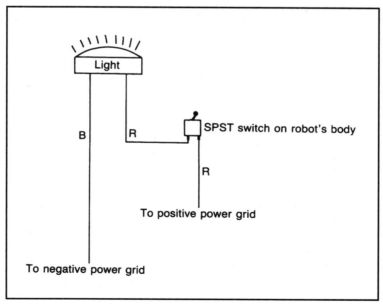

FIGURE 5-18.   Head wiring diagram.

connected. Use the bottom tab and the one marked "open" on the side of the switch. This allows the motor to run only when the door of the vacuum outlet is open. Finally, the fuse for the system is built into the wire itself; be sure the fuse is in place before attempting to use the vacuum. Once wired, all that is needed to activate the system is to open the door to the vacuum outlet and attach the vacuum cleaner hose. This completes all of Questor's major systems.

SKIN AND FINISHING TOUCHES

Now that Questor's major systems have been completed and installed, it is time to improve his appearance. In this chapter you will attach Questor's metal skin, mount the controls, and install other options from previously installed systems that make the robot function and appear as a butler, true to his robot servant theme. Also in this section you'll paint the metal skin and add the robot's serving tray. Before the skin can be attached, Questor's remote control (RC) system, drink dispenser, and arms must be removed. Be sure that you remove the aluminum angles used to mount the RC system and drink dispenser.

SKIN

Questor's skin consists of metal flashing. This material is a very thin sheet metal that can be easily cut with a pair of scissors; the flashing comes in various lengths and widths. You'll need approximately two 8-foot rolls that are 10 inches wide; from this will be cut three 3-foot panels and two 2-foot panels. The two 2-foot panels will be cut in half lengthwise to give you four 5-inch-wide sections. Out of these seven sections will be cut the final skin panels.

Beginning with the three 3-foot panels, trim them so that you have three panels that are 2 1/2 feet long and 9 inches wide. Next cut the four 5-inch-wide panels so that they are 19 inches long. Once all the skin panels have been cut, they are attached to the framework from the inside. You can cut and

TABLE 6-1. Parts List

AMOUNT	ITEM
2	Roll of aluminum flashing
#	Rolls of double-sided tape, 1/2 inch wide
2	Roll of aluminum tape, 6 inches wide
1	2- × 4-foot sheet of hardboard
2	2-inch section of aluminum angle
5	1-inch section of aluminum angle
#	1/8-inch sheet-metal screws
2	Auto courtesy lights
2	SPST switch
1	12-volt horn
1	Rubber gasket (big enough to fit around the vacuum system's motor)
1	4-foot length of rubber floor matting
12	1/8-inch-diameter washer
1	Light-up bow tie
1	9-volt battery
1	Can black spray paint
1	Can white spray paint
1	TV snack tray
1	Roll black foam insulation tape
1	1- × 2-foot sheet of felt

tuck the skin around obstructions along the framework; just be sure it looks good from the outside. The panels are held in place by strips of double-sided tape running the length of the inside of the framework. Also, sheet metal screws along the outside of the framework secure the panels in place. The screws are placed at all the 1/8-inch-diameter predrilled holes on the sides of the framework. Figure 6-1 illustrates where and how the panels are attached. You will need to cut a hole in the center of the lower rear panel to accommodate the vacuum cleaner motor that protrudes there, and a square on the lower front panel to accommodate the vacuum outlet.

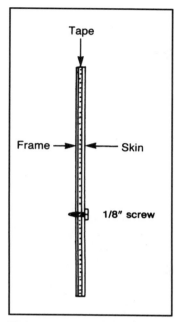

FIGURE 6-1.   Skin attachment to framework.

The metal skin panels cover all of the robot's body except for the rear of the upper framework and the top of the lower framework. These openings will be covered with utility panels which, when removed, allows access to Questor's innerworkings. The utility access panels are made of 1/8-inch-thick hardboard. The panels are then covered with aluminum tape to give them a metallic look that matches the rest of the body.

The panel for the rear of the upper framework measures 9 × 29 inches; it is held to the framework by five 1-inch-long mounting tabs made from leftover aluminum angle used to make the framework. These tabs are riveted to the rear framework and the panel in turn is screwed to the tabs with 1/8-inch sheet metal screws, like those used to secure the metal skin to the framework. Figure 6-2 shows how each tab is to be drilled to accept both the rivet and the screw. Rivet the five tabs along the rear framework where 1/8-inch holes have been drilled previously. Match the 1/8-inch holes on the tab with that on the framework and rivet it in place, as shown in Fig. 6-3.

To locate where on the panel to drill the holes for the screws used to attach it, place the panel on the inside of the

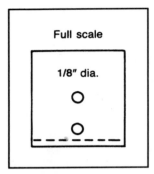

Full scale

1/8″ dia.

FIGURE 6-2.   Mounting tab for rear access panel.

FIGURE 6-3.   Tab in place on framework.

robot's framework so the five tabs sit on top of the panel. Mark the panel where the holes on the tabs rest, then remove it and drill 1/8-inch-diameter holes where marked. Now when you place the panel on top of the tabs, the holes should line up and the panel can be screwed in place. To complete the panel, cover it with aluminum tape. Two 6-inch wide strips running the length of the panel will be enough to cover its front surface and edges. Figure 6-4 shows the completed panel installed.

FIGURE 6-4.   Completed rear access panel.

The access panels that cover the top of Questor's lower framework are made and mounted in the same way as the rear panel. These two U-shaped panels when pieced together form a ring around the upper framework and cover the openings of the lower. The measurements for the panels are shown in Fig. 6-5. The panels are held in place by the same type of metal tab system used for the rear panel. In this case, however, you

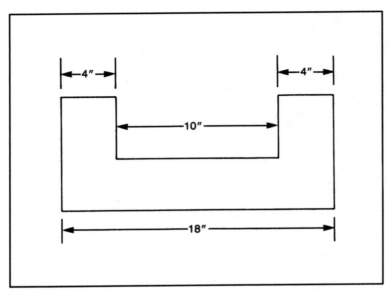

FIGURE 6-5. U-shaped access panel for lower framework.

will need two 2-inch- and two 1-inch-long tabs. The tabs are screwed to the lower framework using 1/8-inch sheet metal screws rather than rivets; I made this decision because I felt it looked better. You, however, could rivet them if you wish. There are predrilled holes along the top of the lower framework where the tabs should be attached. The two 2-inch tabs go on the sides of the frame, while the 1-inch tabs go on the front and rear. Figure 6-6 shows where the tabs are located.

Drilling the mounting holes in both panels and tabs so that they line up, is handled differently from the rear panels. Place the two panels so that they sit flush on the mounting tabs. If the panels go past the inside of the top of the lower framework, trim the edges of the panels until they both fit inside the framework and sit flush on the mounting tabs; now mark on the panels where they sit on the tabs. The two 1-inch tabs will have one hole a piece, while the two 2-inch tabs will have two holes at either end of the two panels where they both sit on the one tab. Figure 6-7 illustrates where 1/8-inch mounting holes will be drilled at the same time in both panels and

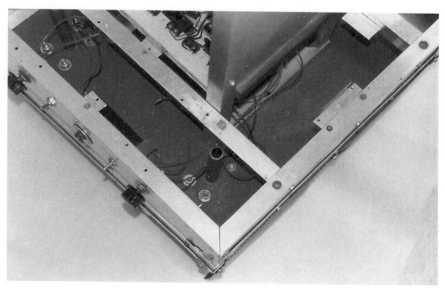

FIGURE 6-6.   Tabs for lower access panels.

the tabs. Cover the panels with aluminum tape and fasten them to the tabs with 1/8-inch sheet metal screws.

Now that all the openings on Questor's body are covered, it is time to mount the robot's serving tray, lower body lights, and controls previously wired to the robot.

MOUNTING TRAY

The next item to be mounted to Questor is his serving tray. If you haven't already, reattach Questor's arm and reinstall his drink dispenser system. The tray itself is a 17 1/2- × 12 3/4-inch snack tray like that pictured in Fig. 6-8. Place the tray across the arms and mark on the tray where the corners sit on the arms. Drill four 1/4-inch-diameter holes in the tray and sit it back on the arms. Align the tray so it sits straight and mark the holes in the tray on the arms. Drill four 1/4-inch-diameter holes in the arms and screw the tray in place with four 1/4-inch-diameter metal screws. Also at this time you'll be attaching Questor's hands that you made previously.

FIGURE 6-7. Hole in both access panel and mounting tab.

MOUNTING CONTROLS

Eight previously installed components are now mounted on Questor's body panels. The first three are the two speed controllings pots and the main power switch of the robot's motorized platform. These components are mounted on the rear panel of the lower framework on either side of the motor for the vacuum system. Figure 6-9 shows where the components are located. The second switch seen on the rear panel is for the two headlights that will be mounted on the front of the lower panel

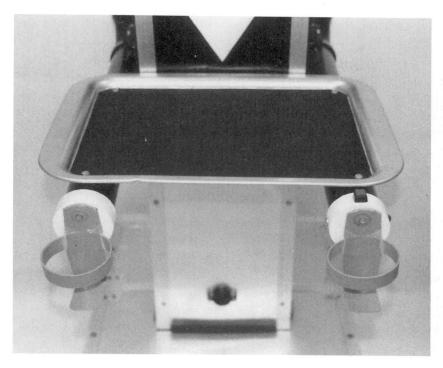

FIGURE 6-8.   Serving tray and hands in place.

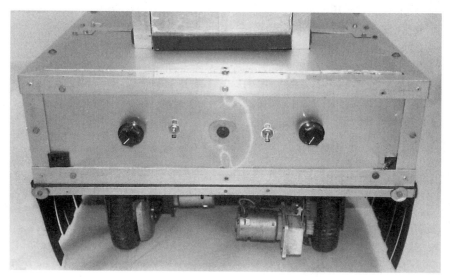

FIGURE 6-9.   Location of components mounted earlier (pots, on/off switch, and charging plugs).

later in this chapter. Also note the rubber gasket around the vacuum cleaner motor that you can attach at this time.

To mount the pots and switches, poke a hole through the front of the panel folding the excess sheet metal against the inside of the panel. It is then a simple matter to mount the components in place by attaching them to the panels using the nuts that come with both the switches and pots. You can, if you wish, buy knobs for the pots, but it is not necessary.

Also mounted on the rear panel of the lower framework are the two charging plugs for Questor's batteries. These plugs needed a firmer base than the metal flashing can provide. For this reason two 6-inch strips of aluminum angle are riveted vertically at either side of the framework behind the body panel. The plugs are then fitted between the side of the framework and the support strip utilizing a ridge that goes around the plug. A small square must be cut in the lower corner on either side of the panel to accommodate the plug as seen in Fig. 6-10.

The next two items are the on/off switch and recharge plugs for the remote control system. The on/off switch is mounted at the lower edge of the right side panel of the upper framework. Follow the mounting directions that come with the RC system and use the hardware that comes with the switch. The RC charging plug is taped to the lower edge of the back framework itself; I used duct tape because it had a metallic look that matched the framework. Both mountings are shown in Figs. 6-11 and 6-12.

The last item to be mounted is the on/off switch for the auto dome light in Questor's head. Mount it on the same side as the on/off switch for the RC system, but at the top edge of the framework, as shown in Fig. 6-13.

Once these eight components have been mounted, Questor is basically complete. The rest of this chapter is devoted to adding items to spruce up the robot's appearance. Also, if you have purchased an RC system with a third channel, you will be adding a horn that is controlled by the system's third servo.

FIGURE 6-10. Charging plug in place.

FIGURE 6-11. RC on/off switch mounting.

FIGURE 6-12.   RC charging plug.

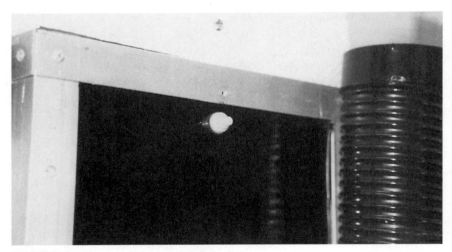

FIGURE 6-13.   On/off button for dome light.

BODY LIGHTS AND HORN

Questor's headlights are two 12-volt auto courtesy lights mounted on the front of the lower framework positioned at either side of the vacuum cleaner port. You mount the lights

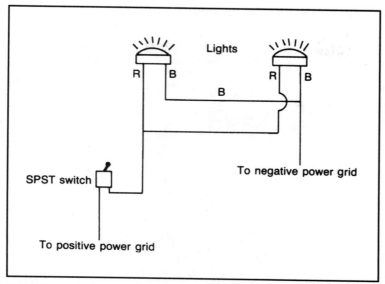

FIGURE 6-14. Courtesy light wiring diagram.

the same way as the switches mounted earlier, but cut out the extra metal instead of folding it in. Follow the mounting directions for the lights you have obtained. Figure 6-14 shows how to wire the lights. The control switch for the lights (an SPST switch you must supply) is mounted on the rear of the lower framework on the opposite side of the vacuum cleaner motor from the main power switch.

Questor's 12-volt horn is mounted on the bottom of the upper framework's front panel; it is then wired to the RC motherboard that, if you opted, should already have the control leaf switch wired in place. Mount the horn on the front of the robot somewhere and connect the red wire of the horn to terminal number 14 on the RC motherboard, and the black lead to the negative power grid. To control the horn, move one of the control sticks on the RC transmitter sideways. This will cause the third servo on the motherboard to turn, activating the leaf switch controlling the horn. A second leaf switch mounted on the opposite side of the servo is used to control any other on/off function you wish. Both the horn and lights are shown in Fig. 6-15 mounted in place.

FIGURE 6-15. Lights and horn in place.

12-VOLT POWER OUTLET

Questor's 12-volt power outlet allows you to draw power directly from his batteries for use with other 12-volt devices, e.g., radios and mini televisions. The outlet itself is a cigarette lighter mount available at most electronics stores. Since most radios and TVs come with auto lighter plugs, this was the logical choice. Mount the outlet on the left side of Questor's lower framework. Figure 6-16 shows how to wire the outlet to the power grid on Questor's remote control board.

BOW TIE

Another light up option you can add to your robot is a bow tie with flashing LEDs (light emitting diodes) like that pictured in Fig. 6-17. The bow tie is available at most novelty shops or from the supplier listed in the back of this book. The bow tie

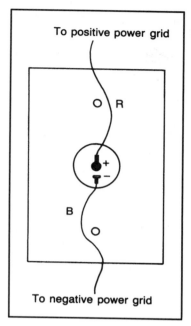

FIGURE 6-16. 12-volt outlet wiring guide.

FIGURE 6-17. Light-up bow tie.

operates off a 9-volt battery taped to the inside of Questor's front body panel. The battery will last for a long time with intermittent use. Also, the tie has no on/off switch, so you'll have to wire and mount one yourself. Figure 6-18 shows how to do this. The tie is held to the robot with a strip of double sided tape like that used to mount the body panels.

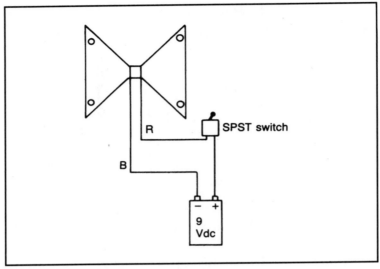

FIGURE 6-18.   Bow tie wiring diagram.

PAINTING AND TRIMMING THE BODY

You can paint your version of Questor any way you wish, but if you are making a duplicate model of my version, the following photos, Figs. 6-19 to 6-22, will better show how he is painted. I only painted the upper portion of the robot so the robot would look as if he were wearing a tuxedo jacket. The two colors used are black and white. To paint the arms it is best to remove them from the body; the upper arm is covered with lengths of leftover black hosing from the vacuum system. Simply cut the hose to length and slide it over the upper arm; the end-cap at the top will hold it in place. Make sure you also paint Questor's serving tray, if needed, to match the rest of his body.

TRIM

There are still a few gaps on Questor's body that need to be covered to enhance his appearance; four of these gaps appear where the panels of the upper framework meet the two access panels at the top of the framework. Cover these gaps using

FIGURE 6-19. Painting guide.

FIGURE 6-20. Painting guide.

FIGURE 6-21.   Painting guide.

black foam weather stripping, as shown in Fig. 6-23. Place the strips so they stick to the upper body panels just touching the lower access panels. When you later remove and replace the access panels, be sure to slide them under the foam strips.

The last item to be mounted to Questor is a skirt to hide his wheels. The skirt is made of black rubber floor matting available at any hardware store. You will need approximately a 2- × 2-foot piece of matting; from this cut four 6- × 20-inch strips, one for each side of the robot's lower body. Cut slits up the matting to allow it to pass over obstacles. Then screw the mat strip directly to the edge of the wooden platform using large washers to help support the strip. Figure 6-24 shows how the installed strips look.

FIGURE 6-22.   Painting guide.

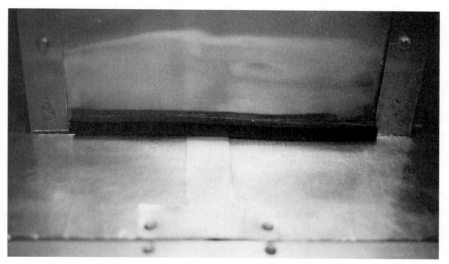

FIGURE 6-23.   Foam strip used to cover body gaps.

FIGURE 6-24.   Platform skirt in place.

Two smaller gaps that need to be covered are found where the two lower access panels on the top of Questor's lower framework meet. These small gaps are covered with T-shaped metal covers made from leftover metal flashing used to make the robot's skin. The strips are held in place by the screws that hold the access panels to its mounting tabs. Figure 6-25 shows what the covers look like and how they are held in place.

The last detail to take care of is to cover the surface of Questor's serving tray with black felt. The felt helps keep items on the tray from shifting when the robot moves. The felt is available with a sticky backing so all you need to do is cut it to fit and stick it in place. Make sure that you obtain felt that is waterproof so the color will not run when it gets wet from drinks carried on the tray. This last little detail completes Questor's construction.

FIGURE 6-25.   T-shaped gap cover for lower access panels.

SOURCES

PARTS SUPPLIERS

Herbach & Rademan Company
353 Crider Avenue
Moorestown, NJ 08057
(800) 848-8001
www.herbach.com

American Science & Surplus
3605 Howard Street
Skokie, IL 60076
(847) 982-0870
www.sciplus.com

The Robot Store
4286 Redwood Highway PMB-N
San Rafael, CA 94903
(800) 374-5764
www.robotstore.com

Radio Shack
300 West Third Street
Suite 1400
Fort Worth, TX 76102
www.radioshack.com

BOOKS

The Robot Builder's Bonanza, 2d ed., by Gordon McComb (McGraw-Hill, 2001)

Robots, Androids, and Animatrons, 2d ed., by John Iovine (McGraw-Hill, 2002)

Personal Robotics: Real Robots to Construct, Program, and Explore the World by Richard Raucci (A.K. Peters, 1999)

Build Your Own Robot! by Karl Lunt (A.K. Peters, 2000)

Robot Riots: The Guide to Bad Bots by Alison Bring and Erin Conley (Dorset Press)

(The following informative but older books are, unfortunately, out of print. However, you should be able to locate them in a library, in a used book store, or on a used book web site.)

The Complete Handbook of Robotics by Edward L. Safford (TAB Books)

How to Build Your Own Self-Programming Robot by David L. Heiserman (TAB Books)

How to Build Your Own Working Robot Pet by Frank DaCosta (TAB Books)

Robot Intelligence...with Experiments by David L. Heiserman (TAB Books)

Robots (Fact, Fiction and Prediction) by Jasia Reichardt (Viking)

Robots Reel to Real by Barbara Krasnoff (Arco Publishing)

The State of the Art Robot Catalog by Phil Berger (Dodd, Mead and Company)

The Robots Are Here by Alvin Silverstein and Virginia B. Silverstein (Prentice-Hall)

MAGAZINE

Robot Science & Technology
3875 Taylor Road, Suite B
Loomis, CA 95650
(888) 510-7728
www.RobotMag.com

SHOWBOT COMPANIES FEATURED

Robots 4 Fun
14807 N. Forestdale Drive
Rathdrum, ID 83858
(208) 687-2923
www.robots4fun.com

The Robot Factory
3740 Interpark Drive
Colorado Springs, CO 80907-5058
(719) 447-0331
www.robotfactory.com

Pelican Beach LLC (successor to ShowAmerica Inc.)
217 Wood Glen Lane
Oak Brook, IL 60523
(630) 530-5673

INDEX